Major Applications of Carbon Nanotube Field-Effect Transistors (CNTFET)

Balwinder Raj
Dr. B. R. Ambedkar National Institute of Technology, Jalandhar, India

Mamta Khosla
Dr. B. R. Ambedkar National Institute of Technology, Jalandhar, India

Amandeep Singh
National Institute of Technology, Srinagar, India

A volume in the Advances in Computer and Electrical Engineering (ACEE) Book Series

IGI Global
DISSEMINATOR OF KNOWLEDGE

Published in the United States of America by
IGI Global
Engineering Science Reference (an imprint of IGI Global)
701 E. Chocolate Avenue
Hershey PA, USA 17033
Tel: 717-533-8845
Fax: 717-533-8661
E-mail: cust@igi-global.com
Web site: http://www.igi-global.com

Library of Congress Cataloging-in-Publication Data

Names: Raj, Balwinder, 1980- editor. | Khosla, Mamta, 1971- editor. | Singh, Amandeep, 1988- editor.
Title: Major applications of carbon nanotube field-effect transistors (CNTFET) / Balwinder Raj, Mamta Khosla, Amandeep Singh, editors.
Description: Hershey, PA : Engineering Science Reference, an imprint of IGI Global, 2019. | Includes bibliographical references and index. | Summary: "This book explores the methods and applications of converting semiconductor devices from micron technology to nanotechnology. It also examines existing CNTs, CNTFETs, and their applications and examines practical applications to minimize short channel effects and power dissipation in nanoscale devices and circuits" -- Provided by publisher.

Identifiers: LCCN 2019029902 (print) | LCCN 2019029903 (ebook) | ISBN 9781799813934 (hardcover) | ISBN 9781799813941 (paperback) | ISBN 9781799813958 (ebook)
Subjects: LCSH: Field-effect transistors--Materials. | Carbon nanotubes.
Classification: LCC TK7871.95 .M284 2019 (print) | LCC TK7871.95 (ebook) | DDC 6213815/28--dc23
LC record available at https://lccn.loc.gov/2019029902
LC ebook record available at https://lccn.loc.gov/2019029903

This book is published in the IGI Global book series Advances in Computer and Electrical Engineering (ACEE) (ISSN: 2327-039X; eISSN: 2327-0403)

British Cataloguing in Publication Data
A Cataloguing in Publication record for this book is available from the British Library.

Advances in Computer and Electrical Engineering (ACEE) Book Series

ISSN:2327-039X
EISSN:2327-0403

Editor-in-Chief: Srikanta Patnaik SOA University, India

MISSION

The fields of computer engineering and electrical engineering encompass a broad range of interdisciplinary topics allowing for expansive research developments across multiple fields. Research in these areas continues to develop and become increasingly important as computer and electrical systems have become an integral part of everyday life.

The **Advances in Computer and Electrical Engineering (ACEE) Book Series** aims to publish research on diverse topics pertaining to computer engineering and electrical engineering. **ACEE** encourages scholarly discourse on the latest applications, tools, and methodologies being implemented in the field for the design and development of computer and electrical systems.

COVERAGE

- Circuit Analysis
- VLSI Design
- Sensor Technologies
- Analog Electronics
- Power Electronics
- Computer Science
- Qualitative Methods
- Computer Architecture
- Chip Design
- VLSI Fabrication

IGI Global is currently accepting manuscripts for publication within this series. To submit a proposal for a volume in this series, please contact our Acquisition Editors at Acquisitions@igi-global.com or visit: http://www.igi-global.com/publish/.

The Advances in Computer and Electrical Engineering (ACEE) Book Series (ISSN 2327-039X) is published by IGI Global, 701 E. Chocolate Avenue, Hershey, PA 17033-1240, USA, www.igi-global.com. This series is composed of titles available for purchase individually; each title is edited to be contextually exclusive from any other title within the series. For pricing and ordering information please visit http://www.igi-global.com/book-series/advances-computer-electrical-engineering/73675. Postmaster: Send all address changes to above address.

Titles in this Series

For a list of additional titles in this series, please visit:
https://www.igi-global.com/book-series/advances-computer-electrical-engineering/73675

Handbook of Research on New Solutions and Technologies in Electrical Distribution Networks
Baseem Khan (Hawassa University, Hawassa, Ethiopia) Hassan Haes Alhelou (Tishreen University, Syria) and Ghassan Hayek (Tishreen University, Syria)
Engineering Science Reference • © 2020 • 370pp • H/C (ISBN: 9781799812302) • US $270.00

Neural Networks for Natural Language Processing
Sumathi S. (St. Joseph's College of Engineering, India) and Janani M. (St. Joseph's College of Engineering, India)
Engineering Science Reference • © 2020 • 227pp • H/C (ISBN: 9781799811596) • US $235.00

Emerging Applications of Differential Equations and Game Theory
Sırma Zeynep Alparslan Gök (Süleyman Demirel University, Turkey) and Duygu Aruğaslan Çinçin (Süleyman Demirel University, Turkey)
Engineering Science Reference • © 2020 • 284pp • H/C (ISBN: 9781799801344) • US $225.00

Handbook of Research on Emerging Applications of Fuzzy Algebraic Structures
Chiranjibe Jana (Vidyasagar University, India) Tapan Senapati (Padima Janakalyan Banipith (H.S.), India) and Madhumangal Pal (Vidyasagar University, India)
Engineering Science Reference • © 2020 • 300pp • H/C (ISBN: 9781799801900) • US $285.00

Electrical Insulation Breakdown and Its Theory, Process, and Prevention Emerging Research and Opportunities
Boxue Du (Tianjin University, China)
Engineering Science Reference • © 2020 • 258pp • H/C (ISBN: 9781522588856) • US $215.00

IGI Global
DISSEMINATOR OF KNOWLEDGE

701 East Chocolate Avenue, Hershey, PA 17033, USA
Tel: 717-533-8845 x100 • Fax: 717-533-8661
E-Mail: cust@igi-global.com • www.igi-global.com

Table of Contents

Detailed Table of Contents

In recent years, carbon nanotube (CNT) emerged as one of the promising materials that shows various advantages over silicon material (e.g., aggressive channel length scaling due to absence of mobility degradation, variable bandgap with single material, ultra-thin body device that is possible due to smaller diameter [1-3nm], and compatibility of CNT with high-k materials resulting in high ION). Moreover, CNTs show both metallic and semiconducting properties; hence, by using metallic CNTs, interconnects can be realized to fabricate a circuit purely consisting of CNTs. This chapter will provide introduction to carbon nanotubes field effect transistor (CNTFETs) starting from material properties of carbon nanotubes and followed by how it can be used as semiconductor channel in field effect transistor (MOSFET) to form CNTFET. The different types of CNTFETs will be discussed based on the type of CNT used along with their advantages and disadvantages.

Section 2
Applications of CNTFETs

The novel characteristics of CNTFET have eliminated many technological and fundamental hindrances being faced by CMOS transistors. CNTFET is emerging as prospective replacement for CMOS transistors in digital circuits and systems. This chapter introduces design of CNTFET-based basic logic gates. The basic logic gates analyzed are inverter, NAND, and NOR gates. The designed gates are evaluated in terms of delay, power consumption, and figure-of-merit power-delay-product (PDP). The standard H-SPICE CNTFET model of Stanford University has been used for all simulations. The impact of dielectric material variations on performance parameters of carbon nanotube field effect transistor based universal gates has been analyzed. Comparison between CMOS and CNTFET-based logic circuits is carried out for different dielectric material at 16 nm technology node.

In the recent digital designs, there are certain circumstances where energy efficiency and ease is required, and in such situations, ternary logic (or three-valued logic) is favored. Ternary logic is an auspicious supernumerary to the conventional binary (0, 1) logic design techniques as this one is possible to attain straightforwardness and energy efficiency. This chapter deals with the comparative analysis of CMOS and CNTFET-based ternary inverter and universal gates design. The simulation result is analyzed and validated with a Hailey simulation program with integrated circuit (HSPICE) simulations. The average delay and power consumption in CNTFET-based ternary inverter have been reduced by approximately 90.3% and 48.8% respectively, as compared to CMOS-based ternary inverter design. Likewise, delay is reduced by 50% and power gets 99% reduction in ternary CNTFET NAND gate as compared to CMOS-based ternary NAND gat. It is concluded that CNFETs are faster and consume less power compared to CMOS technology.

The main objective of this chapter is to provide high-performance, low-power solutions for VLSI system designers. As technology scales down to 32nm and below, the present CMOS technology has to face the scaling limit, such as the increased leakage power, SCEs, and so on. To overcome these limits, the researchers have experimented on other technologies, among which a CNT technology-based device called CNTFET has been evaluated as one of the promising replacements to CMOS technology. In any digital systems, memory is an integral part, and it is also the largest constituent. SRAM is a widely used memory. In today's ICs, SRAM is going to occupy 60-70% of the total chip area. In this connection, this chapter describes the design of CNTFET-based 6T SRAM cell using circuit-level leakage reduction techniques, named sleep transistor, forced stack, data-retention sleep transistor, and stacked sleep.

The chapter will start with brief introduction to the interconnects and its importance in an integrated circuit at deep sub-micron technology nodes. The brief discussion about the concept of scaling, interconnects models, and material in use are presented. The limitations of conventional materials at scaled down technology nodes will be discussed next. The focus of the chapter is to present the electrical equivalent circuit model to estimate the impedance parameters of SWCNT bundle and MWCNT bundle as interconnects at different nano-scaled technology nodes for global level interconnect length. Using ESC model of SWCNT, MWCNT, and copper, the performance comparative analysis for delay and power delay product (PDP) will be presented for different interconnect lengths at nano-scaled technology nodes. Finally, the chapter summary and conclusion will be written at the end of the chapter.

Section 4
Future of CNTFET Technology

Silicene is one of the most interesting nanomaterials. In this chapter, computational studies have been done on Silicene nanotube and nanoribbon-based FETs to analyze their transport properties. The FET is designed from armchair nanoribbon and single wall nanotube. The scattering region is capped by a dielectric and a metallic layer to form a gate. The conductance versus gate bias voltage, conductance versus temperature up to 2000K, and electrode temperature versus current characteristics are calculated and plotted along with the design of the equivalent model of the structure. Extended Huckel-based calculations were used, and the analysis shows the transport properties of both structures.

Preface

CNTs and CNTFET are two important sectors whose applications in VLSI design have huge impact for bringing revolutionary advancements in nanoscale devices, circuits and systems due to improved electronic properties of the CNT nanomaterials. Scaling of conventional MOSFETs towards smaller sizes is near to end because of limited mobility, large Short Channel Effects (SCEs), high leakage currents and power dissipation. To mitigate these challenges of conventional MOSFET devices and to scale the semiconductor device further into nanoscale domain, alternative materials and device structures are required. CNTs and CNTFETs are the potential devices for VLSI circuits and system applications and are highly preferred in novel nanoscale devices due to their high mobility and high driving capacity. Although currently the knowledge and resources for the fabrication of CNTs and CNTFETs are limited but it is predicted by International Technology Road Map for Semiconductors (ITRS) that, with advancement of technology the conventional fabrication flow can be used for the nanoscale devices specifically CNTs and CNTFETs.

The goal of the book *Major Applications of Carbon Nanotube Field-Effect Transistors (CNTFET)* is to provide a platform for domain experts, researchers and practitioners to exchange ideas on technologies and practical experiences related to innovations in current research on CNTFET and its applications.

This book is organized into ten (10) chapters categorized under four (4) different sections depending on the chapter contents.

Section 1 is devoted to providing introduction to CNTFETs. In recent years, CNT emerged as one of the promising material which shows various advantages over silicon material viz. aggressive channel length scaling due to absence of mobility degradation, variable bandgap with single material, ultra-thin body device that is possible due to smaller diameter (1-3nm) and compatibility of CNT with high-k materials resulting in high I_{ON}.

To throw light on the issues related to the use of CNTs in transistors, authors in Chapter 1 have provided introduction to CNTFETs, starting from material properties of carbon nanotubes, followed by how they can be used as semiconductor channel in field effect transistor (MOSFET) to form CNTFET. The different types of CNTFETs are discussed based on the type of CNT used along with their advantages and disadvantages.

In Section 2, few applications based on CNTFETs such as Memory design, Sensor design, Logic gates design are presented along with their performance evaluations and comparison with their CMOS counterparts under Chapters 2-6.

With the scaling of feature size of device, conventional CMOS technology based Static Random Access Memory (SRAM) has reached at its limit due to significant leakage power. Therefore, carbon nanotube field effect transistor (CNTFET) can be considered most suitable alternative for SRAM. Authors in Chapter 2, present the performance and stability analysis of CNTFET based SRAM cell.

Chapter 3 presents designing of CNT based piezoresistive pressure sensors for low pressure range. The design process, mathematical modeling and simulation of CNT based piezoresistive pressure sensors has been covered in details. Also, improved performance of the pressure sensor in terms of linearity and sensitivity in the case of CNT as compared to the silicon counterpart has been demonstrated.

CNTFET is also emerging as prospective replacement for CMOS transistors in digital circuits and systems. Chapter 4 introduces design of CNFET based basic logic gates. The designed gates are evaluated in terms of delay, power consumption, and figure-of-merit power-delay product. Comparison between CMOS and CNFET based logic circuits is also presented.

In the recent digital designs there are certain circumstances where energy efficiency and ease is required, in such situations Ternary logic (or three-valued logic) is favored. Chapter 5 deals with the comparative analysis of CMOS and CNTFET based ternary Inverter and Universal Gates Design. It is concluded that CNFETs is faster and consumes less power as compared to CMOS technology.

Chapter 6 describes the design of CNTFET-based 6T SRAM cell using circuit-level leakage reduction techniques, named as sleep transistor, forced stack, data-retention sleep transistor, and stacked sleep.

A very interesting property of CNTs is that they exhibit both metallic and semiconducting properties, hence by using metallic CNTs, interconnects can be realized to fabricate a circuit purely consisting of CNTs. In recent years, CNT interconnects have emerged as a potential alternative to copper interconnects due to their several magnificent properties. Section 3 coverage is on CNTs as interconnect with three chapters.

Chapter 7 presents a brief introduction to the interconnects and their importance in integrated circuits at deep sub-micron technology nodes. A brief discussion about the interconnects models and materials in use is also presented. The limitations of conventional materials at scaled down technology nodes has been discussed.

In Chapter 8, RLC equivalent model of bundled single-wall CNT is presented by using driver-interconnect-load system with CMOS driver. The crosstalk delay has been studied with the interconnect length, transition time and number of tubes in a bundle.

Chapter 9 presents performance analysis of Mixed-Wall CNT Interconnects using Colliding Bodies Optimization (CBO) Technique. In a MWCB structure, tube density plays a major role to determine the parasitic elements associated with the interconnects. It is inferred from the study that the overall crosstalk noise, delay and power dissipation of MWCB interconnect with higher tube density i.e. obtained using CBO technique are lesser than other CNT structures. The CBO technique proposed in this chapter has been demonstrated to be highly efficient and effective for on-chip interconnects in IC designs.

Section 4 is titled Future of CNTFET Technology and contains one chapter.

In Chapter 10, computational studies have been done on Silicene nanotube and nanoribbon based FETs to analyze their transport properties.

This book presents a comprehensive approach where CNTFETs have been used for VLSI circuits and system applications. The book includes the recent progress and future challenges of this new and exciting field of CNTFETs that are the emerging nanoscale semiconductor devices.

Section 1
Introduction to CNTFETs

Chapter 1
CNTFETs:
Introduction and Types

Amandeep Singh
https://orcid.org/0000-0002-2373-3368
National Institute of Technology, Srinagar, India

Mamta Khosla
National Institute of Technology, Jalandhar, India

Balwinder Raj
National Institute of Technical Teachers Training and Research, Chandigarh, India

ABSTRACT

In recent years, carbon nanotube (CNT) emerged as one of the promising materials that shows various advantages over silicon material (e.g., aggressive channel length scaling due to absence of mobility degradation, variable bandgap with single material, ultra-thin body device that is possible due to smaller diameter [1-3nm], and compatibility of CNT with high-k materials resulting in high ION). Moreover, CNTs show both metallic and semiconducting properties; hence, by using metallic CNTs, interconnects can be realized to fabricate a circuit purely consisting of CNTs. This chapter will provide introduction to carbon nanotubes field effect transistor (CNTFETs) starting from material properties of carbon nanotubes and followed by how it can be used as semiconductor channel in field effect transistor (MOSFET) to form CNTFET. The different types of CNTFETs will be discussed based on the type of CNT used along with their advantages and disadvantages.

DOI: 10.4018/978-1-7998-1393-4.ch001

1. INTRODUCTION

With discovery of CNTs in 1991 (Iijima, 1991), the researchers were interested to know the fundamental properties and technological prospective due to its low dimensional nanostructure. Fundamental properties of CNTs include low dimensionality (Dresselhaus et al., 2007), absence of dangling bonds (Li et al., 2008), near ballistic transport (Rahman et al., 2005), much weaker effect of Fermi pinning (Leonard, Francois & Tersoff, 2000) and exhibiting both metallic & semiconducting properties (Dresselhaus, Dresselhaus & Saito, 1995). The most important characteristic of CNT is that its diameter and bandgap are dependent on its atoms arrangement known as chirality. Hence designers can control the diameter and bandgap by changing the atoms arrangement. CNTs have been used to build prototype nanodevices ranging from carbon nanotube field effect transistor (CNTFET), electromechanical sensors (Mahar et al., 2007) and displays (Choi et al., 1999) in research labs and show outstanding characteristics as compared to conventional devices.

Based on the number of shells present in nanotubes, CNT can be classified into two types i.e. single walled CNT (SWCNT) (McEuen, Paul, Fuhrer & Park, 2002), if there is only single shell; and multi wall CNT (MWCNT) (Li, Hong, Yin, Banerjee & Mao, 2008), if there are more than one shells. The first ever discovered CNT in 1991 was MWCNT but after two years researchers managed to synthesize SWCNT. With discovery of SWCNT, they are used for further development of the CNTFETs as compared to MWCNTs due to better control of bandgap (Ostling, Tomanek & Rosen, 1997).

Among the various applications, this chapter will be mainly focused on the field effect transistor application of CNTs i.e. CNTFETs. Semiconducting SWCNTs act as channel in field effect transistor and depending on the way SWCNT is used/placed, CNTFETs are categorized into four type i.e. Schottky barrier CNTFET (Hazeghi, Arash, Krishnamohan & Wong, 2007), conventional CNTFET (Clifford, John, Castro & Pulfrey, 2004), partially gated CNTFET (Guo et al., 2002) and tunnel CNTFET (Appenzeller, 2005). Schottky barrier CNTFETs consist of undoped CNT that results into Schottky barrier with metals at drain and source sides. Conventional CNTFET consists of doped CNT as channel and work as conventional MOSFET. Partially gated CNTFET consists of partial gate region covering channel, and tunnel CNTFET consists of oppositely doped source and drain regions. This chapter is basically focused on the detailed description of different types of CNTFETs with their advantages and disadvantages for various applications.

1.2 Carbon Nanotubes

The most common form of carbon is graphene which occurs in sheet form with atoms placed in honeycomb nature. CNTs are synthesized by rolling graphene sheet around an axis with some angle. CNTs are hollow cylindrical tubes having diameter in range of nanometer and length upto µm. There are several parameters associated with CNTs which decides its electrical properties. Some parameters are: diameter, circumference, chirality/chiral angel, bandgap, contacts with metal, intrinsic concentration, charge carries for conduction etc. This section discusses in detail the major parameters and characteristics that need to be explored for using CNTs into field effect transistors.

1.2.1 Chirality

While rolling the graphene sheets for synthesizing CNTs, the rolling angle decides the orientation of the axis and is known as chirality vector (Dresselhaus, Dresselhaus, & Saito, 1995). denoted by n and m and represented as (n,m). CNT can be named as chiral, armchair and zigzag based on chirality vectors e.g. $(n,0)$ is named as zigzag and (n,m) is known as chiral and armchair as shown in figure 1.

CNT's conducting properties can be decided using chirality like if (n-m) factor is divisible by 3 (Dresselhaus, Dresselhaus & Saito, 1995), then CNT shows metallic nature else, semiconducting. If probability of synthesizing CNT with all chirality is equal then 1/3 of synthesized CNTs will be metallic and rest 2/3 are semiconducting. Therefore, one can think about more yields for semiconducting nature which is required for field effect transistors at nanoscale regime.

Figure 1. Three types of CNTs

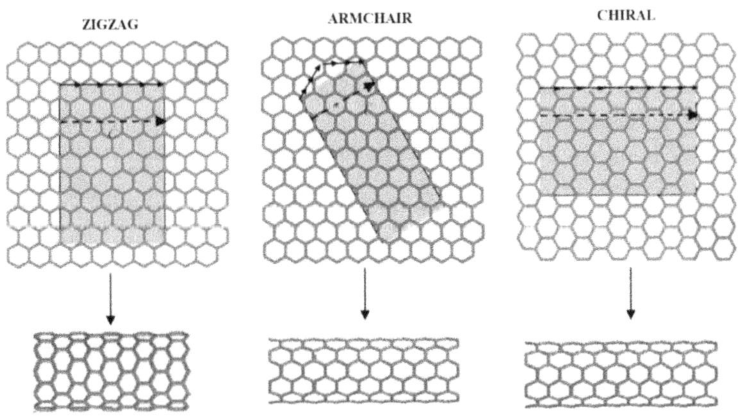

1.2.2 Variable Bandgap

The most important parameter is to find the bandgap of the nanotube synthesized in the process, because it will determine the conduction through it. The bandgap is given by (Dresselhaus, Dresselhaus & Saito, 1995):

$$E_g = \frac{2a\ V_{pp\Pi}}{\sqrt{3}d} \tag{1.1}$$

where d is diameter, $V_{pp\pi}$ is the carbon π–π bond energy and its value is 3.033 eV for tight binding approach and a is inter-atomic distance between carbon atoms and its value is 0.142nm. All the parameters are constant except diameter which is given by (Dresselhaus, Dresselhaus & Saito, 1995):

$$d = \frac{\sqrt{3}a}{\pi}\sqrt{n^2 + m^2 + 2nm} \tag{1.2}$$

It is clear that diameter depends on the chirality vectors only, hence by controlling the orientation of CNT, one can decide its diameter and bandgap. The generalized value of bandgap is given by 0.9/d eV (d is in nanometer), and most common CNT is used as (13,0) having diameter 1nm resulting in bandgap of 0.9 eV which is closed to silicon (1.1eV). This feature of CNTs can bring revolution in semiconductor industry since designers can vary the bandgap of CNTs as they do not have to look for different materials for different bandgap.

1.2.3 CNT- Metal Contact

As our main focus will be field effect transistors based on CNT, the next important characteristic that needs to be explored is CNT- metal contact. Near ohmic contacts are always preferred at source drain junctions to have high ON current. For intrinsic CNT, Schottky contacts are made but contacts can be made near ohmic by using the specific metals as Fermi level pinning is not dominant (Lee, Ung, Gipp & Heller, 2004) and Schottky barriers are solely dependent on work function of metal (Javey et al., 2003). The main reason behind this is dimensionality of the material used and geometry of contact of CNTs made with metal. CNT is a low dimensional material (2-dimensional) (Balleste et al., 2011) as compared to silicon/germanium material and it forms end bonded contact (Zhang et al., 1999) as compared to conventional planar contact in case of silicon-metal contact.

The main reason for Fermi level pinning is metal induced gap states (MIGS) at metal contact which are nothing but electronic states to appear in semiconductor bandgap. For studying the role of Fermi level pinning, the effect of induced charge due to MIGS should be taken into account. In case of end bonded contacts, the MIGS charge takes the shape of dipole ring instead of dipole sheet (Leonard, Francois, & Tersoff, 2000). This shape of dipole has great effect on the electrostatic potential across the CNT and band bending at contacts. In case of dipole sheet the electrostatic potential is constant throughout but in dipole ring it decays third power of distance and hence vanishes in few nanometer from metal contacts. This shows that Fermi level pinning has minor role in end bonded contacts.

The next important aspect is how to achieve the near ohmic contacts by using the metals of specific work function. The barrier at the CNT-metal interface can be calculated as:

$$\varphi_B = \chi\text{m} - \chi\text{s} \tag{1.3}$$

where χ_m is electron affinity of metal and χ_m is electron affinity of CNT. The electron affinity for CNT is around 4.2 eV, and the level of intrinsic Fermi level will be 4.5 eV for CNT of bandgap 0.6 eV i.e. CNT having diameter of 1.6nm (Shiraishi, Masashi & Ata, 2000; Suzuki, Satoru, Bower, Watanabe, & Zhou, 2000). Hence a metal with work function greater than 4.8eV or less than 4.2eV would result into negative Schottky barrier. This would result into near ohmic contact. Metal like Au (5.5 eV) and Pd (5.1eV) can be used to realize ohmic contacts.

1.2.4 Ambipolarity

The main limitation of above method is ambipolar conduction since the CNTs used are intrinsic, hence it will conduct for both bias. Moreover, intrinsic CNTFET exhibits p-type behavior due to the alignment of source-drain metal fermi level near the valance band of the CNTs (Hien, Sy, Luong, Tuan & Nga, 2009). Also, it would be difficult to achieve the near ohmic contacts for CNTs with smaller diameter or large bandgap since metal with higher work function will be requirement. It has been reported that zero Schottky barrier can be achieved for smallest diameter to be 1.5nm (Kim et al., 2005). In order to overcome these limitations the method of doping CNTs is used which will be helpful for suppressing ambipolar current and attaining high value of unipolar current. This is due to the fact that doping CNTs will create excess of majority carriers and the current will be only due to majority carriers as in case of conventional MOSFET.

The doped CNTs form the ohmic contact with metal due to Fermi level alignment with little effect of work function of metal. However doping CNTs is not easy as conventional doping techniques are not applicable to them (Lin, Appenzeller, Knoch & Avouris, 2005). The reason behind this is a dopant atom can destroy the desired arrangement of carbon atoms that defines chirality and hence the desired semiconducting properties can be disturbed. Chemical doping techniques are deployed in case of CNTs which shows excellent properties as compared to conventional silicon MOSFET (Javey et al., 2005). A lot of work has been done to investigate the optimum doping level in CNTs for best ON current (I_{ON}), OFF current (I_{OFF}) and I_{ON}/I_{OFF} (Alam, Khairul, & Lake, 2007). It has been observed that optimum doping level is $5x10^{-3}$ dopants per atom which shows I_{ON} value 10μA and I_{OFF} value 1nA giving I_{ON}/I_{OFF} around 10^4 (Alam, Khairul, & Lake, 2007). If the doping level is $1x10^{-2}$ dopants per atom, both I_{ON} and I_{OFF} value increase but I_{OFF} increases with greater order than I_{ON} which reduces I_{ON}/I_{OFF} (Alam, Khairul, & Lake, 2007).

1.3 CNTFETS and Its Types

CNTFETs are simply MOSFETs with semiconductor material or in other words channel is made up of CNTs. Here the classification is made on the basis of CNT doping i.e. doped CNT or intrinsic CNT. Based on the type of CNT used, CNT can be categorized into four types (Singh, Khosla & Raj, 2015): Schottky barrier CNTFET (SBCNTFET), partially gated CNTFET, conventional CNTFET, and Tunnel CNTFET as shown in figure 1.2.

1.3.1 Schottky Barrier CNTFET (SBCNTFET)

As the name suggests, Schottky barrier CNTFET consists of Schottky contacts with metals at source and drain sides. Schottky contacts are due to the intrinsic type of CNT as shown in figure 2 which will create mismatch between valence/conduction band and Fermi level of metal can be visualized in figure 3.

This mismatch will create a barrier called Schottky barrier which reduces the ON current. Reducing Schottky barrier will enhance ON current however reducing barrier height for n-type will increase barrier for p-type since Fermi level of metal aligns to middle of bandgap for intrinsic CNT. This will result into ambipolar characteristics due to significant contribution from both electrons and holes, making it difficult to turn off the device. Also due to barrier, the conduction takes place due to tunneling instead of thermionic emission, hence the thermal limit of inverse subthreshold slope (60 mV/dec) cannot be achieved in this type of device. Due to above mentioned disadvantages; SBCNTFET is least preferred type of CNTFET.

Figure 2. Structure of SBCNTFET

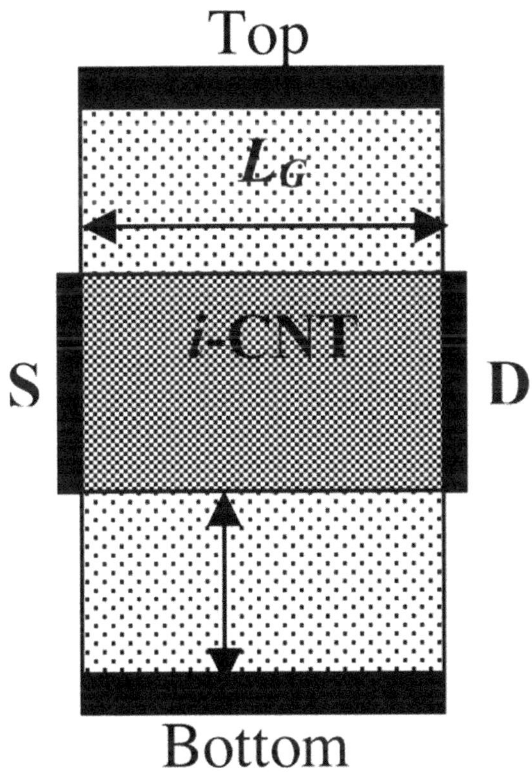

Figure 3. Band diagram of SBCNTFET

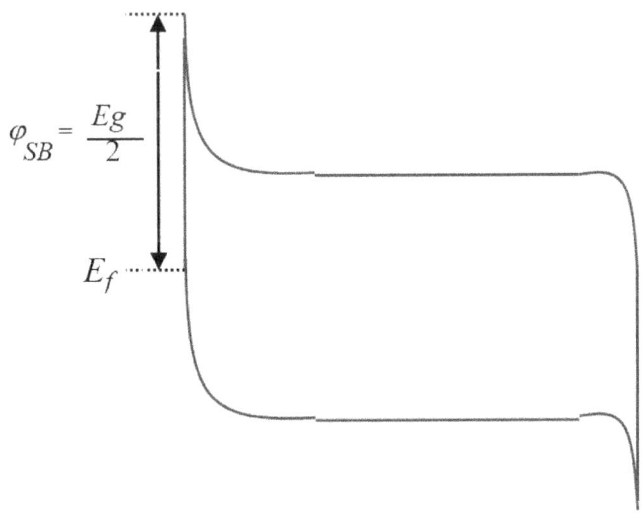

Figure 3 shows the band diagram of SBCNTFET which clearly shows the Schottky barriers at metal contact which is formed due to mismatch of the fermi level of metal and conduction band of CNT. In this case the electrons from the source side have to cross this barrier before going to channel. Similarly, the holes have to cross the barrier to go to channel from drain side. This will result in ambipolar currents. The crossing of barrier can also result in tunneling of carriers through the barrier.

1.3.2 Conventional CNTFET

It is similar to the conventional type MOSFET with doping at source and drain sides as shown in figure 4. With introduction of doping, Fermi level of source and drain will no longer remain at middle of bandgap which will remove barrier and result in ohmic contact with metal. Conventional CNTFET shows unipolar characteristics and due to thermoionic emission the inverse subthreshold slope is attainable to theoretical limit of 60mV/dec, hence it is easy to turn off device with very small OFF current. However, the level of doping should be 10^{-3} dopants per atom (Alam, Khairul, & Lake, 2007) for optimal I_{ON}/I_{OFF} ratio. For higher doping ratios ($5x10^{-3}$), the drain to source tunneling current will dominate the leakage current and it will reduce the I_{ON}/I_{OFF} ratio with several orders of magnitude. With the listed advantages, conventional CNTFET is a better choice as compared to SBCNTFET. However, the conventional CNTFET suffers from problem of charge pile up (Appenzeller, 2005) which consists of injection of charge from drain to gate region affecting moving the bands to turn the device OFF. Figure 5 shows the band diagram of conventional CNTFET. It consist of ohmic contacts are source and drain sides, hence no barriers and the flow is only controlled with gate barrier. As gate potential will increase the barrier will go down providing smooth flow of electrons from source to drain.

1.3.3 Partially Gated CNTFET

This type of CNTFET consists of partial gate metal cover over the CNT channel as shown in figure 6. In this, CNT is either intrinsic or uniformly doped throughout. If intrinsic CNT is used, it will results into Schottky barrier with metal contacts but the effect of Schottky barrier is suppressed as gate effect is not involved at source and drain junctions. Here I_{ON} is dependent on Schottky barrier on the source side since it is not modulated by gate. If doping is done then contacts are ohmic and it will work in depletion mode due to doped charge carriers. Here, it is hard to turn off the device due to throughout doping inside the channel too. Hence, leakage current is a big problem of partially gated CNTFET.

Figure 4. Structure of conventional CNTFET

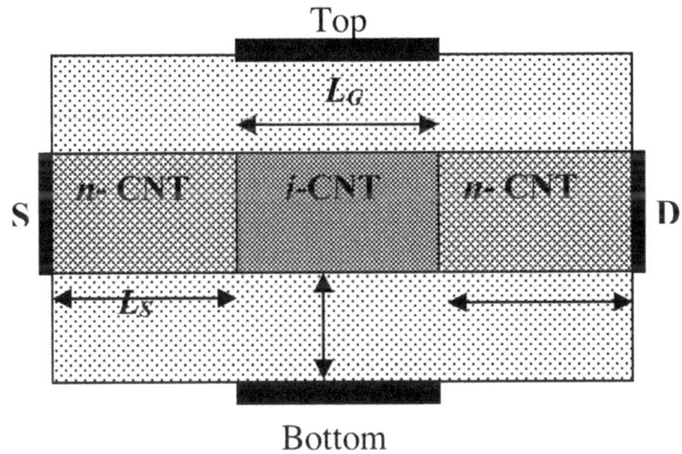

Figure 5. Band diagram of conventional CNTFET

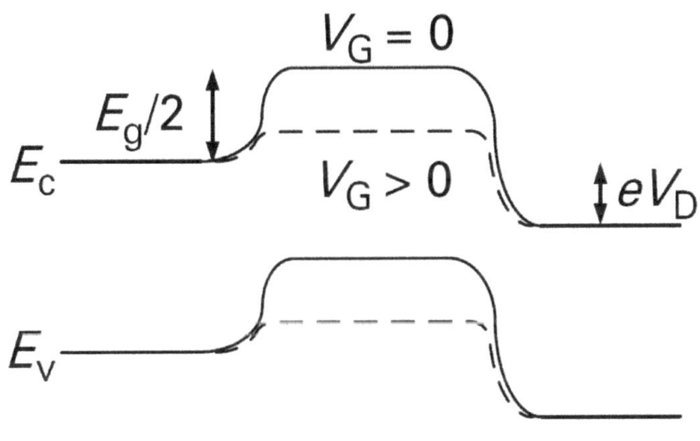

1.3.4 Tunnel CNTFET

Tunnel CNTFET consists of *p-i-n* or *n-i-p* doping profile along the CNT channel as shown in figure 7. This type of CNTFET works on the principle of tunneling mechanism (Knoch & Appenzeller, 2008). The main benefit of tunnel CNTFET is that the inverse subthreshold slope can be achieved below thermal limit of 60mV/dec. It is due to the phenomenon of band to band tunneling in tunnel CNTFET as compared to conventional CNTFET where switching mechanism depends on carrier injection from thermally broadened Fermi function i.e. thermionic emission. Hence

Figure 6. Structure of partial gated CNTFET

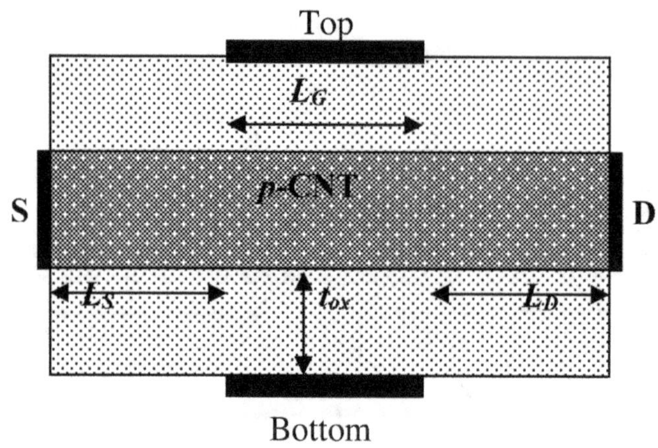

Figure 7. Structure of tunnel CNTFET

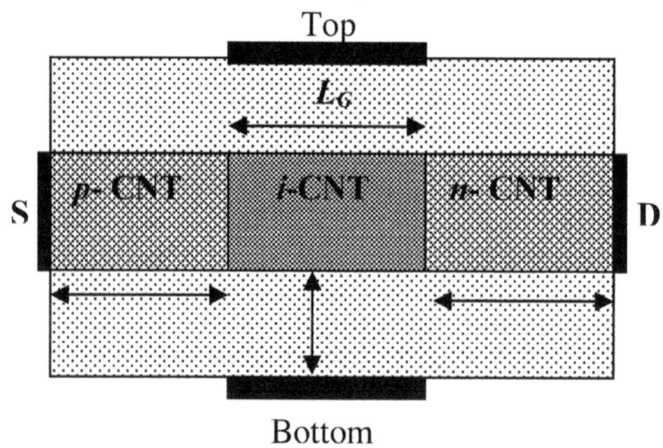

it is easy to turn off a tunnel CNTFET with very small OFF current. Moreover, the problem of charge piles up is not present in tunnel CNTFET which makes them most suitable for circuit applications.

1.3.5 Comparison of Various Types of CNTFETs

Out of the four types of CNTFETs discussed, the most popular are SBCNTFET and conventional CNTFET. However, in recent times tunnel CNTFET gets great attention due to ambipolarity in SBCNTFET and charge pile up in conventional type. The charge piles up can be defined as the stuck of carriers inside channel region while injected from drain. This problem is more severe in ultra short channel device which makes it a leaky transistor. Also, this charge piles up results into large value of subthreshold slope in short channel device.

As discussed above, the common problem in both SBCNTFET and conventional CNTFET is the large value of subthreshold slope. The reason behind this is the mechanism of carrier injection inside the channel i.e. thermionic emission of carriers from source side. If the carrier injection is replaced by band to band tunneling then this problem will be solved, resulting into new device i.e. tunnel CNTFET. Figure 8 shows the band diagram of tunnel CNTFET which clearly shows the tunneling of electrons from conduction band of source to valance band of channel regions. The electrons within energetic window $\Delta\emptyset$ will contribute towards the drain current. The tunneling probability can be made large by reducing tunneling width λ, resulting into large drain current. The main advantages of tunnel CNTFET are very low subthreshold slope and low off currents. However, the major disadvantage is low ON current which is important for digital applications.

Irrespective of the geometry type, CNTFETs mostly conduct through ballistic transport which can be determined by comparing channel length and mean free path. CNTs have mean free path of the order of μm, hence the CNTFETs with channel length of order of nm accounts for ballistic transport. In ballistic transport, the carriers do not suffer through scattering inside the channel and hence have greater mobility leading to higher ON current which makes it good candidate as compared to silicon based MOSFETs.

Figure 8. Band diagram of tunnel CNTFET

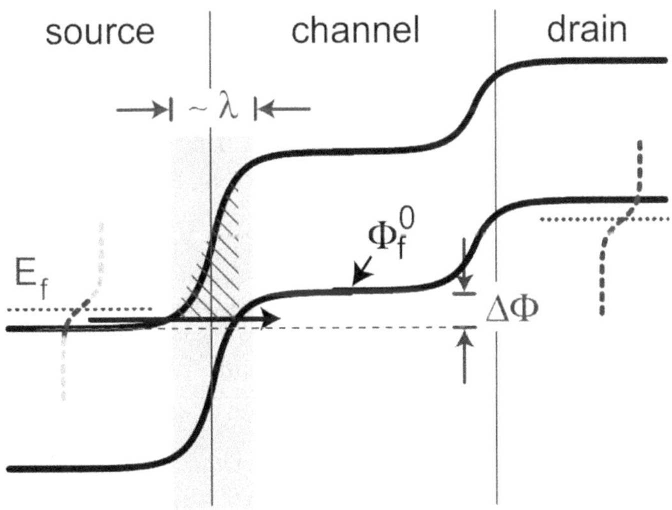

REFERENCES

Alam, K., & Lake, R. (2007). Role of doping in carbon nanotube transistors with source/drain underlaps. *IEEE Transactions on Nanotechnology*, *6*(6), 652–658. doi:10.1109/TNANO.2007.908170

Appenzeller, J., Lin, Y.-M., Knoch, J., Chen, Z., & Avouris, P. (2005). Comparing Carbon Nanotube Transistors—The Ideal Choice: A Novel Tunneling Device Design. *IEEE Transactions on Electron Devices*, *52*(12), 2568–2576. doi:10.1109/TED.2005.859654

Choi, W. B., Chung, D. S., Kang, J. H., Kim, H. Y., Jin, Y. W., Han, I. T., ... Kim, J. M. (1999). Fully sealed, high-brightness carbon-nanotube field-emission display. *Applied Physics Letters*, *75*(20), 3129–3131. doi:10.1063/1.125253

Clifford, J. P., John, D. L., Castro, L. C., & Pulfrey, D. L. (2004). Electrostatics of partially gated carbon nanotube FETs. *IEEE Transactions on Nanotechnology*, *3*(2), 281–286. doi:10.1109/TNANO.2004.828539

Dresselhaus, M. S., Chen, G., Tang, M. Y., Yang, R. G., Lee, H., Wang, D. Z., ... Gogna, P. (2007). New directions for low-dimensional thermoelectric materials. *Advanced Materials*, *19*(8), 1043–1053. doi:10.1002/adma.200600527

Dresselhaus, M. S., Dresselhaus, G., & Saito, R. (1995). Physics of carbon nanotubes. *Carbon*, *33*(7), 883–891. doi:10.1016/0008-6223(95)00017-8

Guo, J., Datta, S., Lundstrom, M., Brink, M., McEuen, P., Javey, A., . . . McIntyre, P. (2002). Assessment of silicon MOS and carbon nanotube FET performance limits using a general theory of ballistic transistors. *Digest International Electron Devices Meeting, IEEE*, 711-714.

Hazeghi, A., Krishnamohan, T., & Philip Wong, H.-S. (2007). Schottky-barrier carbon nanotube field-effect transistor modeling. *IEEE Transactions on Electron Devices*, *54*(3), 439–445. doi:10.1109/TED.2006.890384

Hien, D. S., Luong, N. T., Thi, T. A. T., & Nga, D. V. (2009). 3D simulation of coaxial carbon nanotube field effect transistor. *Journal of Physics: Conference Series*, *187*, 012061. doi:10.1088/1742-6596/187/1/012061

Iijima, S. (1991). Helical microtubules of graphitic carbon. *Nature*, *354*(6348), 56–58. doi:10.1038/354056a0

Javey, A., Guo, J., Wang, Q., Lundstrom, M., & Dai, H. (2003). Ballistic carbon nanotube field-effect transistors. *Nature*, *424*(6949), 654–657. doi:10.1038/nature01797 PMID:12904787

Javey, A., Tu, R., Farmer, D., Guo, J., Gordon, R., & Dai, H. (2005). High performance n-type carbon nanotube field-effect transistors with chemically doped contacts. *Nano Letters*, *5*(2), 345–348. doi:10.1021/nl047931j PMID:15794623

Kim, W., Javey, A., Tu, R., Cao, J., Wang, Q., & Dai, H. (2005). Electrical contacts to carbon nanotubes down to 1 nm in diameter. *Applied Physics Letters*, *87*(17), 173101. doi:10.1063/1.2108127

Knoch, J., & Appenzeller, J. (2008). Tunneling phenomena in carbon nanotube field-effect transistors. *Physica Status Solidi. A, Applications and Materials Science*, *205*(4), 679–694. doi:10.1002/pssa.200723528

Lee, J. U., Gipp, P. P., & Heller, C. M. (2004). Carbon nanotube p-n junction diodes. *Applied Physics Letters*, *85*(1), 145–147. doi:10.1063/1.1769595

Leonard, F., & Tersoff, J. (2000). Role of Fermi-level pinning in nanotube Schottky diodes. *Physical Review Letters*, *84*(20), 4693–4696. doi:10.1103/PhysRevLett.84.4693 PMID:10990773

Leonard, F., & Tersoff, J. (2000). Role of Fermi-level pinning in nanotube Schottky diodes. *Physical Review Letters*, *84*(20), 4693–4696. doi:10.1103/PhysRevLett.84.4693 PMID:10990773

Li, H., Yin, W.-Y., Banerjee, K., & Mao, J.-F. (2008). Circuit modeling and performance analysis of multi-walled carbon nanotube interconnects. *IEEE Transactions on Electron Devices*, *55*(6), 1328–1337. doi:10.1109/TED.2008.922855

Li, X., Wang, X., Zhang, L., Lee, S., & Dai, H. (2008). Chemically derived, ultrasmooth graphene nanoribbon semiconductors. *Science*, *319*(5867), 1229–1232. doi:10.1126cience.1150878 PMID:18218865

Lin, Y.-M., Appenzeller, J., Knoch, J., & Avouris, P. (2005). High performance carbon nanotube field-effect transistor with tunable polarities. *IEEE Transactions on Nanotechnology*, *4*(5), 481–489. doi:10.1109/TNANO.2005.851427

Mahar, B., Laslau, C., Yip, R., & Sun, Y. (2007). Development of carbon nanotube-based sensors—A review. *IEEE Sensors Journal*, *7*(2), 266–284. doi:10.1109/JSEN.2006.886863

Mas-Ballcstc, R., Gomcz-Navarro, C., Gomcz-Ilcrrcro, J., & Zamora, F. (2011). 2D materials: To graphene and beyond. *Nanoscale*, *3*(1), 20–30. doi:10.1039/C0NR00323A PMID:20844797

McEuen, P. L., Fuhrer, M. S., & Park, H. (2002). Single-walled carbon nanotube electronics. *IEEE Transactions on Nanotechnology*, *99*(1), 78–85. doi:10.1109/TNANO.2002.1005429

Ostling, D., Tomanek, D., & Rosen, A. (1997). Electronic structure of single-wall, multiwall, and filled carbon nanotubes. *Physical Review B: Condensed Matter and Materials Physics*, *55*(20), 13980–13988. doi:10.1103/PhysRevB.55.13980

Rahman, A., Guo, J., Datta, S., & Lundstrom, M. S. (2003). Theory of ballistic nanotransistors. *IEEE Transactions on Electron Devices*, *50*(9), 1853–1864. doi:10.1109/TED.2003.815366

Shiraishi & Ata. (2000). Work function of carbon nanotubes. *MRS Online Proceedings Library Archive*, 633.

Singh, M. K., & Raj, B. (2015). Comparative analysis of carbon nanotube field effect transistors. *Proceedings of the IEEE 4th Global Conference on Consumer Electronics*, 552-555. 10.1109/GCCE.2015.7398601

Suzuki, S., Bower, C., Watanabe, Y., & Zhou, O. (2000). Work functions and valence band states of pristine and Cs-intercalated single-walled carbon nanotube bundles. *Applied Physics Letters*, *76*(26), 4007–4009. doi:10.1063/1.126849

Zhang, Y., Ichihashi, T., Landree, E., Nihey, F., & Iijima, S. (1999). Heterostructures of single-walled carbon nanotubes and carbide nanorods. *Science*, *285*(5434), 1719–1722. doi:10.1126cience.285.5434.1719 PMID:10481005

Section 2
Applications of CNTFETs

Chapter 2
CNTFET-Based Memory Design

Shashi Bala
Chandigarh Group of Colleges, Landran, India

Mamta Khosla
National Institute of technology, Jalandhar, India

Raj Kumar
University Institute of Engineering and Technology, Panjab University, Chandigarh, India

ABSTRACT

As the feature size of device has been scaling down for many decades, conventional CMOS technology-based static random access memory (SRAM) has reached its limit due to significant leakage power. Therefore, carbon nanotube field effect transistor (CNTFET) can be considered most suitable alternative for SRAM. In this chapter, the performance and stability of CNTFET-based SRAM cells have been analyzed. Numerous figures of merit (FOM) (e.g., read/write noise margin, power dissipation, and read/write delay) have been considered to analyze the performance of CNTFET-based. The static power consumption in CNTFET-based SRAM cell was compared with conventional complementary metal oxide semiconductor (CMOS)-based SRAM cell. Conventional CNTFET and tunnel CNTFET-based SRAMs have also been considered for comparison. From the simulation results, it is observed that tunnel CNTFET SRAM cells have shown improved FOM over conventional CNTFET 6T SRAM cells without losing stability.

DOI: 10.4018/978-1-7998-1393-4.ch002

1. INTRODUCTION

Over a few decades, as the integrated circuit (IC) technology has been growing rapidly, the feature size of transistors has dropped from micrometer to nanometer regime for having high packaging density on a single chip and increasing processing speed. However, the incessant demand for placing more number of CMOS devices on single-chip causes to reduce the device dimensions, leading to disturbing the switching characteristics of the transistors. Though, feature size of the transistor can be scaled down to nanometre regime by using scaling rules to minimize these disturbances. Subthreshold characteristics of CMOS devices degrade severely as moved to nanometre regime, leading to putting a limit on further scaling of conventional MOS transistors. But, consistent research and development are required for further scaling of technology to follow the International Technology Roadmap for Semiconductors (ITRS, 2013; Young, 1989). Therefore, it becomes essential to explore some alternative non-conventional CMOS devices. There are various nanoscale transistors reported by researches like Tunnel Field Effect Transistors (TFET) (Seabaugh & Zhang, 2010; Kim, 2010) impact ionization metal oxide semiconductor (I-MOS) (Gopalakrishnan, Griffin, & Plummer, 2002; Gopalakrishnan, Woo, Jungemann, Griffin & Plummer, 2002), FinFET (Raj, Saxena & Dasgupta, 2009), Nanowire FET (Scarlet, Ambika & Srinivasan, 2019; Sharma, Raj & Khosla, 2017), Nanotube FET (Tekleab, Tran, Sleight & Hidambarro, 2012; Tekleab, 2014), Carbon nanotube field-effect transistor (CNTFET), Tunnel CNTFET (Knock & Appenzeller, 2008; Shaker, Ossaimee, Zekry & Abouelatta, 2015) etc could be the solution for future CMOS applications.

2. CARBON NANOTUBE FIELD-EFFECT TRANSISTOR (CNTFET)

Carbon nanotube transistors may be considered as promising candidates for FETs due to exceptional electronic, mechanical and thermal properties of carbon nanotubes (CNTs) (Nardelli, Fattebert, Orlikowski, Roland, & Zhao, 2000; Avci, Rios, Kuhn & Young, 2011). CNTFET is having a similar device structure as Si-based MOSFET except for channel material, which is to be replaced by CNT of few nanometer diameters. Variable bandgap attributed to variable diameter of CNTFET makes it useful for different applications. In conventional MOSET, different materials have to be selected for different bandgaps. CNTFET requires low gate voltage as compared to conventional MOSFET (Pourfath, Mahdi, Kosina, & Selberherr, 2007; Koswatta, Nikonov, & Lundstrom, 2005). Therefore, CNTFETs are considered a highly promising candidate for future low power applications. According to their geometry CNTFET is of two types: planer and coaxial. Figure 1 (a) shows the planer CNTFET structure in which source and drain are connected through CNT that is utilized for channel instead of bulk silicon of substrate (Knoch & Appenzeller, 2008). Since CNT is cylindrical nanotube, coaxial geometry of CNTFET is also possible as shown in Figure 1(b). The planer CNTFETs are more popular and easily fabricated than coaxial attributed to their simplicity and compatibility with the manufacturing technology. The selection of geometry is based on its application (Auth & Plummer, 2008).

Single-wall Carbon nanotube (SWCNT) made up of only one carbon sheet rolled in cylindrical form, considered as the most suitable type for using as a channel in the transistor. An SWCNT can behave as a conductor or a semiconductor, depending upon chirality. The chiral vector is given by C=n.a1+m.a2. Where *(n, m)* is chiral number. The diameter of the CNT is depended on chiral number can be calculated as

$$d = \frac{\sqrt{3}a}{\pi}\sqrt{n^2 + m^2 + 2nm} \tag{1}$$

The bandgap of the nanotube (E_g) is inversely proportional to the diameter of the tube as given in (2)

$$E_g = \frac{2E_p a}{d} \tag{2}$$

Where a is the interatomic distance between the atoms, a=0.142nm, E_p gives the carbon π–π bond energy is equal to 3.033eV for tight binding approach and d is the diameter of a nanotube is given in (1)

The minimum voltage required to switch on a transistor is called the threshold voltage (V_{th}). V_{th} of the intrinsic CNT channel may be given by the half bandgap that is an inverse function of the diameter as in (3)

Figure 1. CNTFET structure for (a) planer (b) coaxial geometry (Knoch &Appenzeller, 2008)

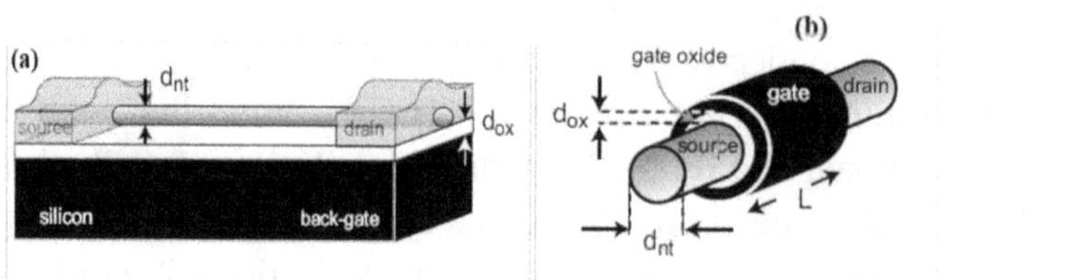

$$V_{th} = \frac{E_g}{2e} = \frac{\sqrt{3}}{3}\frac{aE_p}{ed_{cnt}} \tag{3}$$

Where, e is the charge of an electron.

The drain current (I_{DS}) (Bala, & Khosla, 2018) of CNTFET can be calculated as

$$I_{DS} = \left(\frac{4qKT}{h}\right)\sum\left[\ln\left(1+\exp\left(\frac{E_{FS}-E_c}{KT}\right)\right)-\ln\left(1+\exp\left(\frac{E_{FD}-E_c}{KT}\right)\right)\right] \tag{4}$$

Where, h is the plank's constant and q is the charge of an electron. E_{FS}, E_{FD} are the Fermi energies at the source and drain. K is Boltzmann constant and T is temperature.

3. STATIC RANDOM ACCESS MEMORY (SRAM)

Among embedded memories, SRAM has been continuing to play a key role in every VLSI system attributed to its tremendous speed and compatibility with logic process technology. SRAM widely used as cache memory in microprocessors and application-specific integrated circuits can occupy about 70% portion of the die area (Yamauchi, 2007; 2008). As the memory technologies are entering the nanoscale regime, they are bound to encounter challenging issues of their continued scaling. Since SRAM is created from CMOS technology, several issues related to MOSFET technology scaling i.e. signal loss due to exponentially increase in leakage current, degradation of noise margin cause to V_{TH} mismatch. The reliability issue is also a major concern due to process variability and Random dopant Fluctuation (RDF) (Cheng et al., 2004). To achieve low-voltage operation in SRAM, several challenges have to be faced such as originating from process variation, related to bitcell read /write stability and sensing ability. Figure 2 illustrated the schematic diagram of a standard 6T SRAM cell. 6T SRAM cell is constructed from two cross-coupled inverter pairs, designed in such a way that output node of one inverter pair is connected to input node of second inverter pair. The inverter pairs M1, M2, and M3, M4 need to be further connected to two access transistors M5 and M6. The read/write operation can be performed by applying enough voltages to word lines (WL) and bit lines (**BL**) of access transistors. The gate terminals of access transistors required to be connected to WL and drain terminals of access transistors are connected to bit lines (BL and BLB). At nodes Q and QB original and complementary data values are stored.

Figure 2. Schematic diagram of a standard 6T SRAM cell (Singh, Jawar, Mohanty, & Pradhan, 2012)

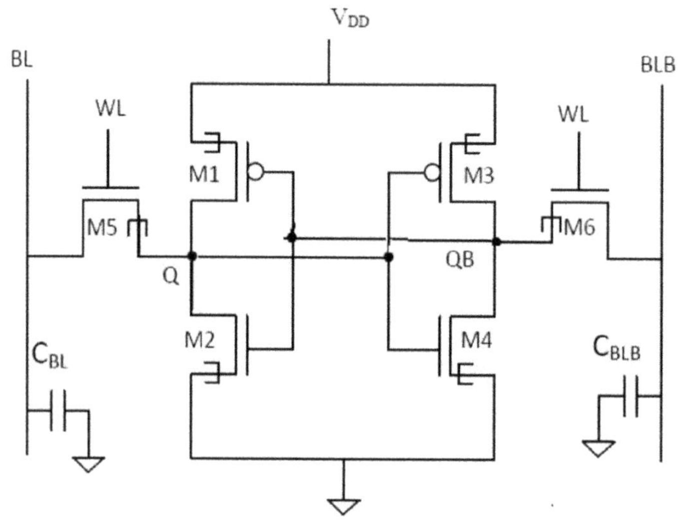

For a read operation, both the bit lines (BL and BLB) are precharged to supply voltage i.e. V_{DD} and word line need to be activated. If the value stored at Q is1 then it will stay at 1 because of the absence of discharge path. If the value 0 is stored at node (Q) then it switches to intermediate voltage instantly, because of current path exists between bit line and ground. The intermediate voltage is governed by the voltage dividers which are constructed by using one an access transistor and a pull-down transistor. The intermediate voltage should not be crossed the threshold voltage of inverter else unwanted voltage flips at internal nodes. Therefore, the pull-down transistor should be stronger than the access transistor for successful read operation. Thus, cell ratio (β) should be high. The difference between bit line (BL) and Q can be sensed by the sense amplifier and indicates that read 0 operation has been done. The read current path in 6T standard SRAM cell is shown in Figure 3.

$$\beta = \frac{\left(\frac{W}{L}\right) ratio\ of\ N\ /\ P\ transistor}{\left(\frac{W}{L}\right) ratio\ of\ access\ transistor} \quad 5$$

The write operation can be performed through the pull-up transistor M1 and the access transistor M5. The voltage levels are asserted to the bit lines according to the data to be written over. For writing 1, BL is asserted to 1 and BLB is asserted

Figure 3. Read current path

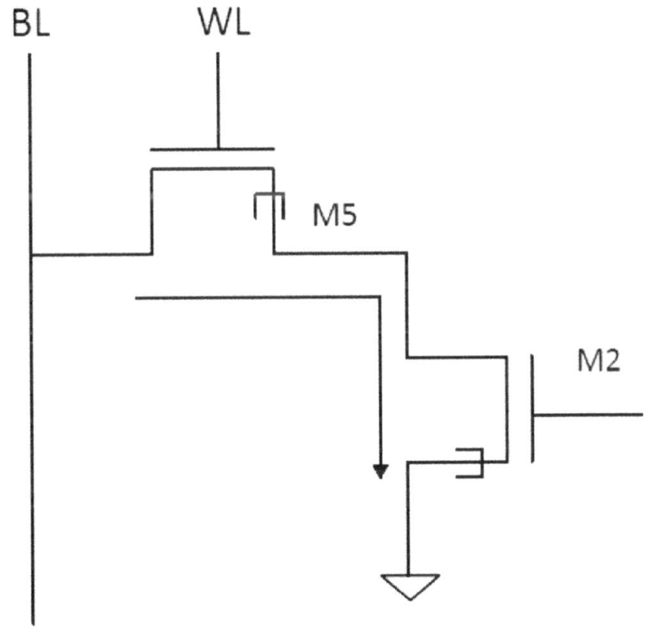

with a value 0 and word line is also activated. Now, the storage node will jump to an intermediate node, decided by pull up transistor M1 and access transistor M5. The storage voltage should be greater than the threshold voltage otherwise read circuit will be activated by the inverter pair. So the access transistor is made stronger to overcome this issue. The write current path in 6T standard SRAM cell is shown in Figure 4.

4. PERFORMANCE PARAMETERS OF SRAM

The read margin, write margin and hold margin is the performance parameter of SRAM. These are briefly described below.

4.1 Hold Noise Margin

Hold noise margin (HNM) is defined as the maximum amount of noise voltage that is barred by inverter pairs so that the bit cells are able to hold the data. HNM is calculated by fitting the biggest square in the hold butterfly curve.

Figure 4. Write current path

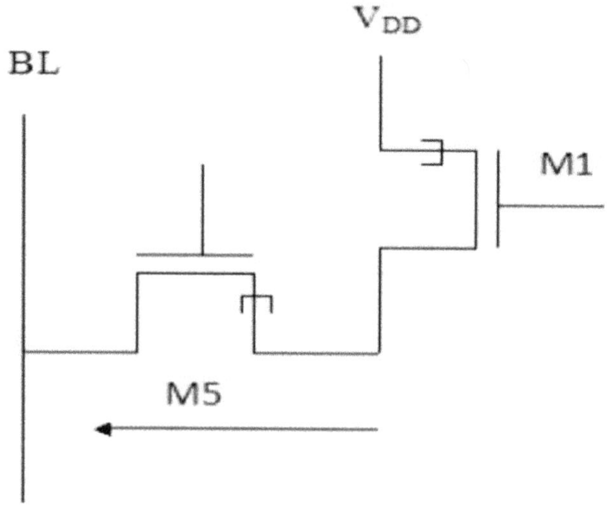

4.2 Read Noise Margin

Read noise margin (RNM) is calculated using the read voltage transfer characteristic which is measured by sweeping the DC voltage at node Q and supervise the voltage at node QB. It is also calculated by fitting the biggest square in the reading butterfly curve.

4.3 Write Noise Margin

Write noise margin (WNM) is defined as the minimal amount of bit line voltages that are required to flip the internal nodes of the cell. WNM is calculated by fitting the biggest square in the read curve and write curve.

Figure 5 shows that the read SNM of 6T SRAM cell is gradually decreasing with technology scaling, while the leakage current is exponentially increasing. Figure 6 represents the V_{DD} scaling with technology length variations in recent years. This shows that after scaling of device length up to 32 nm the V_{DD} limited at 1.0 V. The scaling limit of the transistor is one of the primary reasons to become SRAM V_{DD} scaling so challenging. The increasing V_{TH} random variation can degrade SRAM functional margins, leading to an increase of the minimum SRAM V_{DD} (Yamauchi, 2007).

Figure 5. Comparison of normalized read Static Noise Margin (SNM) and leakage current of a 6T SRAM bit-cell for different technology nodes (Singh, Jawar, Mohanty, & Pradhan, 2012)

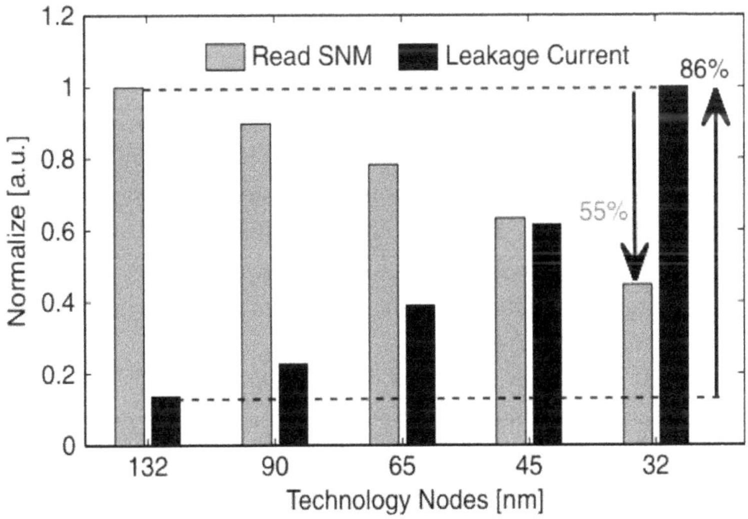

Figure 6. SRAM-operating voltage scaling trend (Yamauchi, 2007)

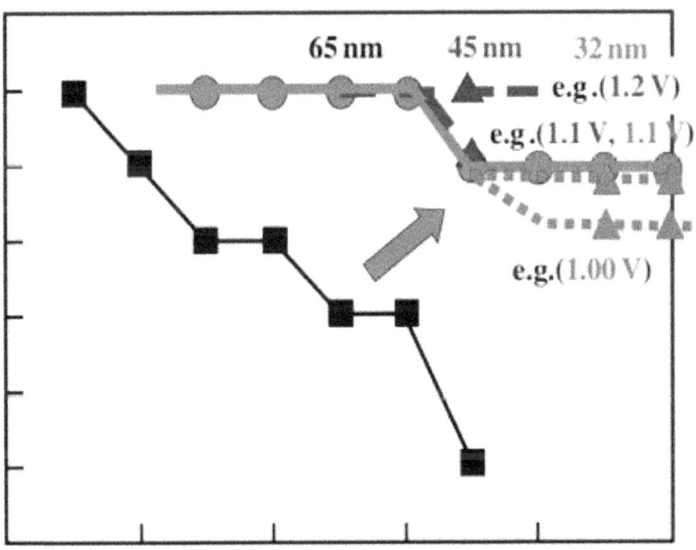

5. CNTFET BASED DIFFERENT SRAM CELL DESIGN

5.1 6T CNTFET SRAM Cell

Figure 7 illustrated the schematic diagram of CNTFET based 6T SRAM cell. CNTFET 6T SRAM cell is the same as the stranded SRAM cell the only difference the MOS transistors replace with CNT transistors. Working of conventional MOSFET based 6T SRAM cell and CNTFET based 6T SRAM have the same is explained in section 3.

CNTFET based SRAM CNTFETs is having a steep subthreshold transistor of extremely low leakage at V_{DD} for a low power and high speed embedded memory applications. CNTFETs have the ability to save critical leakage power for SRAM. Table I summarizes the comparison of the CMOS 6T SRAM cell and CNTFET 6T SRAM cell at 0.9V power supply.

CNTFET based SRAM design has shown tremendous improvements in stability, the sensitivity of process variations and performance, power consumption at a low power supply as shown in Table 1.

Figure 7. Schematic diagram of a CNTFET based 6T SRAM cell (Bala & Khosla, 2018)

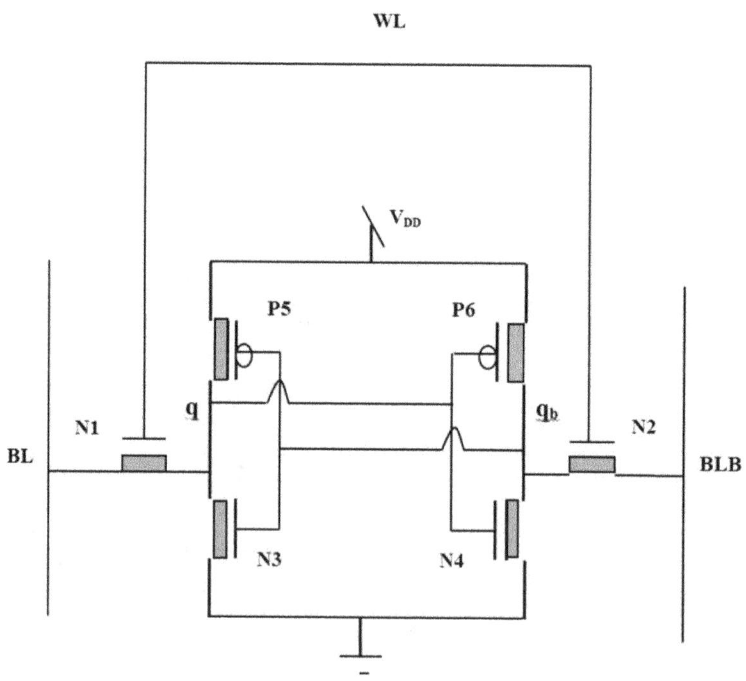

Table 1. Comparison of 6T SRAM 32nm CMOS and 20nm CNTFET at 0.9V (Marani & Perri, 2019)

Feature	CNTFET Lg=20nm	CMOS Lg=32nm
SNM(volts)	0.227	0.085
Access Time (ps)	31.59	25.79
Write time(ps)	29.75	25.62
Power standby(pw)	1.753	81.62

A conventional SRAM uses six transistors to store each memory bit. However, some other kinds of SRAM chips use 5Transistors, 7 Transistors and 8 Transistors per bit.

5.2 5T CNTFET SRAM Cell

In a conventional 6T cell both storage nodes are accessed by NMOS pass-transistors. This is mandatory to write on the cell because internal cell nodes cannot be pulled up from a stored '0' by a high on the bit line. Otherwise, accidental write could occur while reading a stored '0'. On the other side, if the bit lines are not precharged to V_{CC} this is no longer true. The cell can be constructed with an intermediate precharge voltage, $V_{PC.}$ Thereby, a high on the bit line could write a '1' into the cell. However, a precharged bit line with a lower voltage would not be able to do the same. a low on the bit line could also be written a '0' into the cell, whereas the intermediate precharge voltage could not be. Thus a precharge voltage window needs to provide to the cell for ensuring correct operation. The need for two NMOS transistors is eliminated because the cell can be written both high and low from one side. In turn, one-bit line also deducted. This is very attractive from a high-density point of view. Figure 8 shows the structure of the proposed 5T SRAM cell. 5T cell shares a sense amplifier between two cells, with one less bit line. Thereby, reduction in area gives an even greater advantage to 5T memory block over the 6T SRAM.

Writing in the 5T SRAM cell is done by only one bit as compared to the 6T cell. The value to be written is held on the bit line and the word line is asserted for the 5T cell. The reading operation of the 5T cell is quite similar to the 6T SRAM. The word line needs to be held low (grounded) and bit line is precharged before the read operation. However, the bit line is not precharged to V_{CC}, but to another value, V_{PC} this time. The value V_{PC} needs to be chosen carefully as per stability and performance requirements. There is one drawback of the intermediate precharge

Figure 8. Schematic diagram of a CNTFET based 5T SRAM cell (Bhat, 2012)

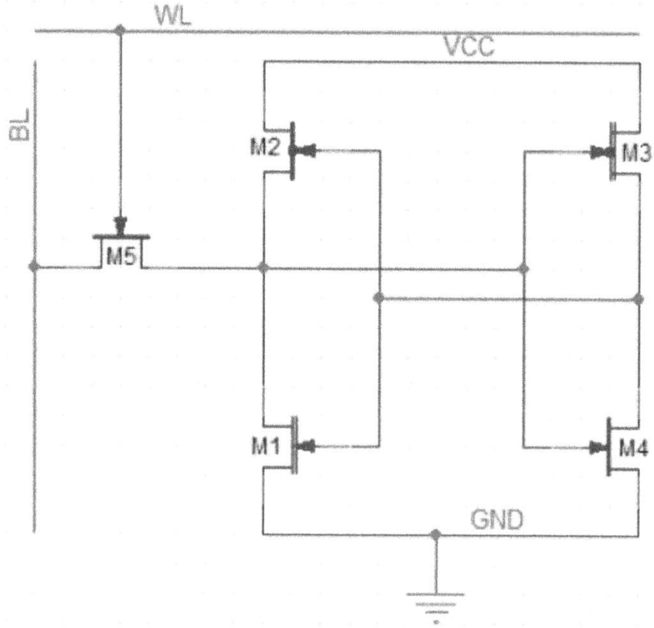

value is evident problem of obtaining this voltage. Nowadays, microprocessors require several different supply voltages, so this might not be a significant drawback.

5.3 7T CNTFET SRAM Cell

7T-SRAM cell based on CNTFETs has been designed to show improvement in the read cycle and that reduces the dynamic power. The schematic of the cell is shown in Figure 9. In the feedback loop, a transistor M7 is added and creates a separate read line 'Read Bit' from the word line 'Write Bit' of the 6-T SRAM cell.

The four transistors M1, M2 and M3, M4 are placed at the center form two cross-coupled inverters INV1 and INV2. A low input on the first inverter INV1 will generate a high on the second inverter INV2 due to the feedback structure, which amplifies and stores the low value on the second inverter INV2. Likewise, a high input on the first inverter INV1 will generate a low input on the second inverter INV2, which feeds back the high input onto the first inverter INV1. Thereby, the two inverters INV1 and INV2 will be able to store their current logical value. But, the feedback circuit connection is established by adding an extra MOS transistor M7. The data is stored at a node 'Q' and its complement will be stored at a node 'Q bar'. The two separate transistors M5 and M6 are used in the circuit to write and read the data from memory cells. To write data into the cell 'Write Select' signal is used and 'Read Select' signal is used to read the data from the cell.

This proposed 7T CNTFET SRAM cell depends on the feedback connection between the two inverters, INV1 and INV2, before a write operation. The feedback connection and disconnection is done by an extra n MOS transistor M7. M7 will be OFF during write operation and ON for reading operation. The cell relies on 'Write Bit' to perform a write operation.

To start write operation, M7 must be turned off to cut off the feedback connection. It allows the fast transfer of the logic value from the write bit line 'Write Bit' into the memory. M5 can be turned on by using a signal 'Write Select', while M6 must be kept off. The 7T SRAM cell appears as two cascaded inverters, INV1 followed by INV2. M5 transistor will transfer the data from 'Write Bit' to Q1 which further drives INV1, M1, and M2, to develop 'Q bar'. Similarly, 'Q bar' drives INV2, M3, and M4, to develop 'Q', the cell data. Then, M5 will get turned off and M7 is turned on to reconnect the feedback link between the two inverters to store the new data. As the switching activity during memory accesses reduces, dynamic power reduces. The 'Write Bit' line needs not to be pre-charged for the read operation and a write operation can affect only a single bit line of the cell as compared to the 6-T memory cell for both operations.

A transistor M6 needs to turn on for read operation using a signal 'Read Select' and turn off the transistor M5. The feedback path is established by turning on 'Write Bar' signal during the operation. Then, the stored data can be read by 'Read Bit' at a node 'Q'. The read cycle can be enhanced by two aspects of the cell operations, one is the ability to pre-charge the read bit line 'Read Bit' irrespective of the activity

Figure 9. Schematic diagram of a CNTFET based 7T SRAM cell (Bhat, 2012)

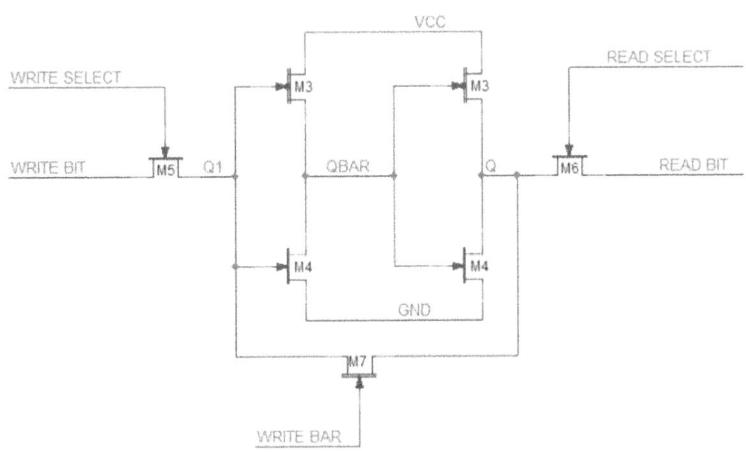

of the write bit line 'Write Bit' and device sizing of the read zero through the pull-down transistor M3 of the second inverter, which made it 8 times larger than the transistor M6 to provide a fast forward path to ground.

5.4 8T CNTFET SRAM Cell

From figure 10, 8T SRAM cell uses a buffered read to isolate the internal nodes of the cell from the read path. The read bit line RBL should be precharged to Vdd before the read operation. RBL will be either remaining at Vdd (if internal node "q" contains a "0") or will be pulled down to ground (if internal node "q" contains a "1"). The internal nodes remain undisturbed for both cases. The bit lines are recharged to the pre-determined values before the write operation. The write operation is started by asserting the WWL and the nodes attain the respective values from the bit lines.

5.5 9T CNTFET SRAM Cell

The CNTFET based 9T SRAM cell realized with cross-coupled inverters created by the transistors L1, D1, L2, and D2, are used to store a single bit of information, as shown in Figure 11. The write bit line WBL and the pass transistor A2 plays key role to transfer new data into the cell. However, the read bit line RBL and transistors E2, E3 and E4 can be used for data reading from the cell.

5.6 10T CNTFET SRAM Cell

The CNTFET based 10T SRAM bit cell is designed by a fully differential read sensing scheme, as shown in Figure 12. In the read mode, WL needs to be enabled

Figure 10. Schematic diagram of a CNTFET based 8T SRAM cell (Joshi & Alabawi, 2017)

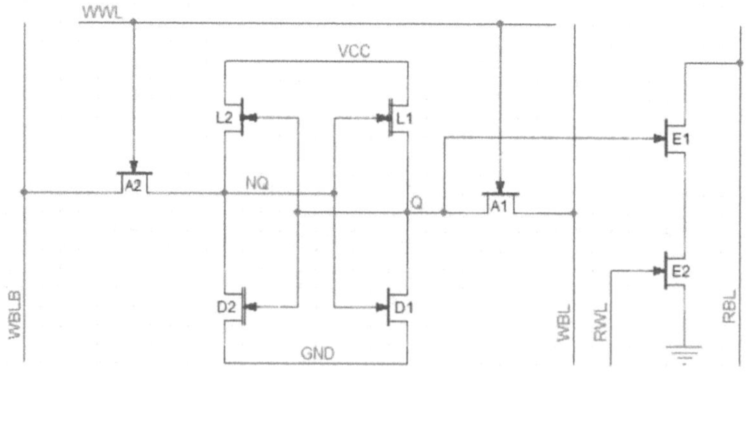

Figure 11. Schematic diagram of a CNTFET based 9T SRAM cell (Joshi & Alabawi, 2017)

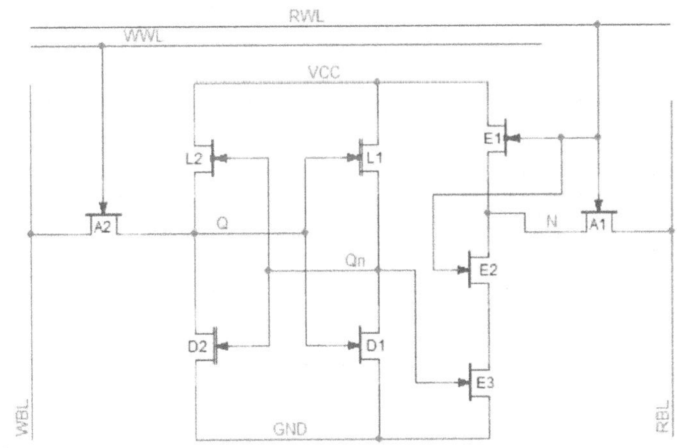

and Vgnd is forced to 0 V, while WWL will remain disabled. The data nodes Q and QB decoupled from the bit line by disabling WWL, during the read access. This isolation keeps the read SNM of the 10T SRAM cell almost same as that of the hold SNM of the 6T SRAM cell. One of the bit lines gets discharged, based on the cell data value once WL is enabled. It has been found that the read value is developed as an inverted signal of cell data in 10T SRAM cell. The bit lines BL and BLB are recharged to the pre-determined values before the write operation. For write mode, both the word lines WL and WWL need to be enabled to transfer the write data to the cell nodes from the bit lines. The writing ability may be a critical matter since 10T SRAM cell has series access transistors.

6. TUNNEL CNTFET

Figure 13 shows the schematic diagram of the 6T SRAM cell using Tunnel CNTFET that has been adopted from (Singh, Ramakrishnan, Mookerjea, Datta, Vijavkrishnan & Pradhan, 2010) by replacing all the Si-based TFETs with Tunnel CNTFET. It consists of both inward and outward access transistors. Inward access configuration is adopted to obtain acceptable read noise margin of the device whereas outward access configuration is used to increase the write noise margin of the device. The inward access transistor couples the internal node Q to BL and outward access transistor couples internal node Q to BLB. The write enables the WR_A signal is used to provide virtual grounding to an inverter pair for obtaining acceptable write noise margin.

Figure 12. Schematic diagram of a CNTFET based 10T SRAM cell (Joshi & Alabawi, 2017)

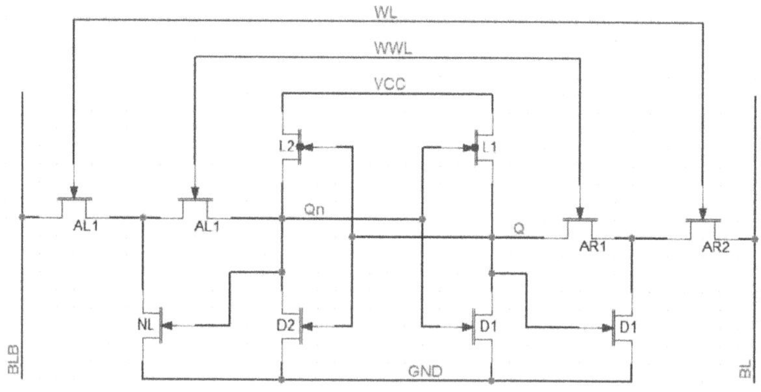

Table 2. Comparison of parameters at different technology nodes and different design approach

Parameter	Technology						
	CMOS 6T SRAM (20 nm) [36]	**CNTFET 6T SRAM (20 nm) [36]**	**CNTFET 7T SRAM (20 nm) [32]**	**CNTFET 8T SRAM (20 nm)[33]**	**CNTFET 9T SRAM (16 nm) [34]**	**CMOS 9T SRAM (65 nm) [35]**	**CNTFET 10T SRAM (20 nm) [32]**
Average Read Power	N.A	N.A	N.A	31.2 (μW)	28 (μW)	160(μW)	N.A
Average Write Power	N.A	N.A	N.A	40.52 (μW)	27.27 (μW)	54.18(μW)	N.A
Leakage Power	33(pW)	8(pW)	4.87(nW)	6.52 (nW)	1.78 (nW)	N.A	17.58(nW)
Read delay	8.9(pS)	2.7(pS)	3.6(nS)	118.21(pS)	.092 (nS)	102(nS)	8.0(nS)
Write delay	9.5(pS)	3.2(pS)	N.A	14.0(pS)	64(nS)	78(nS)	N.A
Read Noise Margin	55(mV)	80(mV)	223(μV)	301(mV)	N.A	N.A	432(μV)
Write Noise Margin	181(mV)	220(mV)	360(μV)	175(mV)	N.A	N.A	475(μV)

Read operation of 6T Tunnel CNTFET SRAM cell is similar to 6T standard SRAM cell. During read operation WR_A is driven to the ground. For writing 1 onto internal node Q, BL and BLB are raised to V_{DD} and then word line is activated. To weaken the first inverter WR_A is raised simultaneously as well. WR_A signal is sent in the form of short pulse whose magnitude is less than the supply voltage but greater than the ground voltage. When Q settles to 1 and QB settles to 0 then WR_A is linked to ground line which enables cross-coupling between two signals. 6T-SRAM cell is designed and simulated in HSPICE for Tunnel CNTFET. The conventional CNTFET 6T SRAM cell is designed using model file of Stanford University (Stanford University Nanoelectronics Group, 2018). Tunnel CNTFET based SRAM cell is designed using model (Bala & Khosla, 2018) and shows improvement over CNTFET based SRAM in terms of power dissipation, read/write delay and SNMs without losing stability. Table 3 shows the comparison of Tunnel CNTFET SRAM with conventional CNTFET SRAM at 0.5V. The calculated values read noise margin, write noise margin and static power dissipation are improved in Tunnel CNTFET SRAM cell. The power reduces from 9pW to 6pW at 0.9V. Therefore the power reduction will be significant for large memory applications. Hence it can be determined that Tunnel CNTFET SRAM is suitable for low power applications.

Figure 14 and Figure 15 shows read noise margin and write noise margin at various V_{DD} and it is seen that on increasing V_{DD}, the Tunnel CNTFET shows improvement in read margin and write noise. It is visualized that the write noise margin improved more than 20% and this improvement increases with higher V_{DD}.

Figure 13. Schematic diagram of a Tunnel CNTFET based 6T SRAM cell

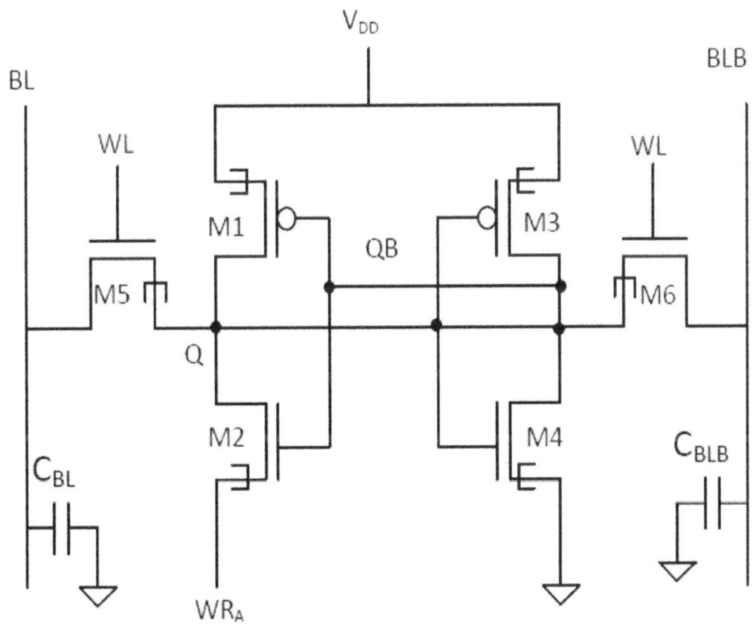

Table 3. Comparison of Tunnel CNTFET and Conventional CNTFET SRAM at V_{DD}=0.5V

Parameters/Device	Conventional CNTFE 6T SRAM[38]	Tunnel CNTFET 6T SRAM
Read SNM (mV)	83	87
Write SNM (mV)	220	240
Static power (pW)	9	6

Figure 14. Read noise margins of Tunnel CNTFET 6T SRAM at different V_{DD}

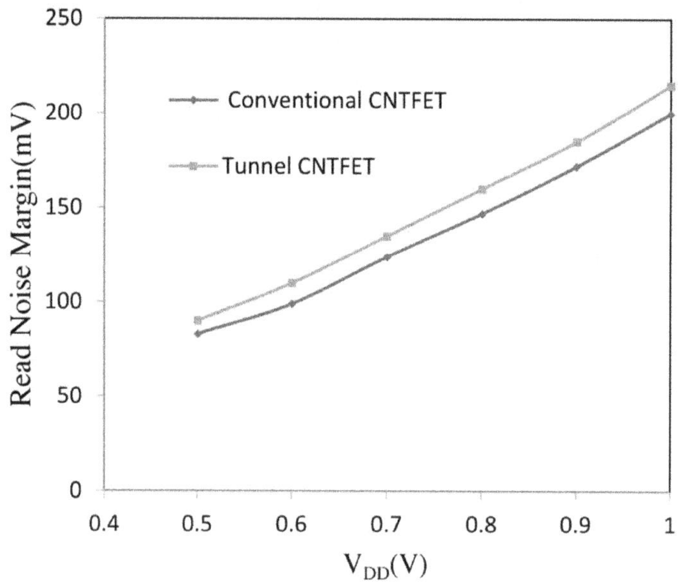

7. CONCLUSION

Memory components have been playing a very significant role in digital electronics for many decades. SRAM is considered as one of the most popular memory elements due to its tremendous ability to retain data, used in most of the digital electronics systems. As the technology scaling down, the design of SRAM has become severe challenge in terms of delay, noise margin, and stability. This chapter illustrates the performance comparison of various CNTFET based SRAM cell topologies like 6T, 7T, 8T, 9T, and 10T cell with respect to read noise margin (SNM), write noise margin (WNM), read delay, and power consumption. The power consumption of 7T SRAM structure found to be minimum at 4.87 nW and is 36% less than 6T structure. The dynamic power increases with around 36%, 67%, and 75% as compared to 6T

Figure 15. Write noise margins of Tunnel CNTFET 6T SRAM at different V_{DD}

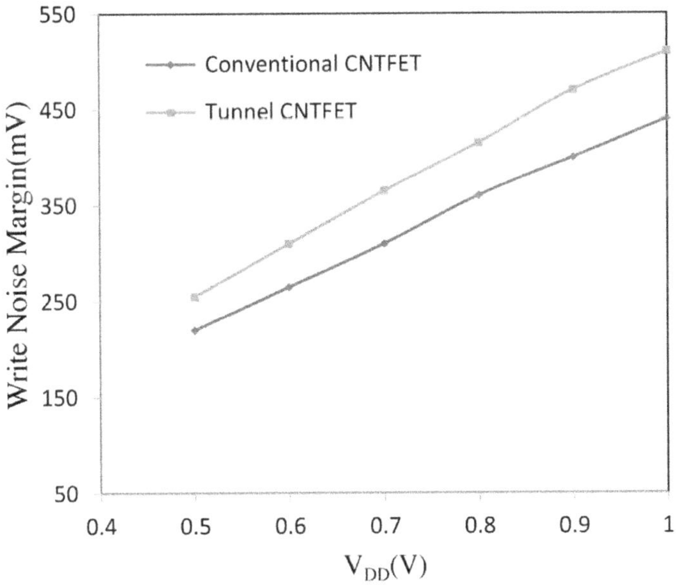

structure for 8T, 9T, and 10T structure. Similarly, the read delay has been noted least for 7T structure i.e. 3.54 psec. which is increased by around 68%, 73%, 74%, and 76% in 6T, 8T, 9T, and 10T, respectively, as compared to 7T structure. Further conventional CNTFET based 6T SRAM are compared with Tunnel CNTFET based 6T SRAM. Simulation results show that Tunnel CNTFET SRAM cell has higher read and write noise margin and moreover it consumes less power, which makes Tunnel CNTFET an appropriate applicant for low power SRAM design.

REFERENCES

Auth, Ch., & Plummer, J. D. (1997). Scaling theory for cylindrical, fully depleted, surrounding gate MOSFET's. *IEEE Electron Device Letters*, *18*(2), 74–76. doi:10.1109/55.553049

Avci, U. E., Rios, R., Kuhn, K., & Young, I. A. (2011). Comparison of performance, switching energy and process variations for the TFET and MOSFET in logic. *2011 Symposium on VLSI Technology-Digest of Technical Papers*, 124-125.

Bala, S., & Khosla, M. (2018). Comparative Study and Analysis of CNTFET and Tunnel CNTFET. *Journal of Nanoelectronics and Optoelectronics*, *13*(3), 324–330. doi:10.1166/jno.2018.2234

Bala, S., & Khosla, M. (2018, September). Electrostatically doped tunnel CNTFET model for low-power VLSI circuit design. *Journal of Computational Electronics*, *5*, 1–8.

Bala, S., & Khosla, M. (2018). Analysis of CNTFET for SRAM Cell Design. *Nanoscale Devices: Physics, Modeling, and Their Application*, 205.

Bhat, N. S. (2012). Design and modelling of different SRAM's based on CNTFET 32nm technology. *International Journal of VLSI Design & Communication Systems* *3*(12).

Boucart, K., & Ionescu, A. M. (2007). Length scaling of the double gate tunnel FET with a high-k gate dielectric. *Solid-State Electronics*, *51*(11), 1500–1507. doi:10.1016/j.sse.2007.09.014

Cheng. (2004). The impact of random doping effects on CMOS SRAM cell. *Proc. ESSCIRC*, 219–222.

Choi, W., & Park, J. (2016). A charge-recycling assist technique for reliable and low power SRAM Design. *IEEE Transactions on Circuits and Systems*, *63*(8), 1164–1175. doi:10.1109/TCSI.2016.2589118

Colinge, J. P. (2008). The new generation of SOI MOSFETs. *Romanian Journal of Information Science and Technology*, *11*(1), 3–15.

Gopalakrishnan, K., Griffin, P. B., & Plummer, J. D. (2002). I-MOS: a novel semiconductor device with a subthreshold slope lower than kT/q. *International Electron Devices Meeting*, 289-292. 10.1109/IEDM.2002.1175835

Gopalakrishnan, K., Woo, R., Jungemann, C., Griffin, P. B., & Plummer, J. D. (2005). "Impact ionization MOS (I-MOS)" Part II: Experimental results. *IEEE Transactions on Electron Devices*, *52*(1), 77–84. doi:10.1109/TED.2004.841344

Hirlekar, Yamagar, Garse, Vij, & Kadam. (2009). Carbon nanotubes and its applications: a review. *Asian Journal of Pharmaceutical and Clinical Research*, *2*(4), 17-27.

Joshi, S., & Alabawi, U. (2017). Comparative Analysis of 6T, 7T, 8T, 9T, and 10T Realistic CNTFET Based SRAM. *Journal of Nanotechnology*, *2017*, 1–9. doi:10.1155/2017/4575013

Kim, Y. B. (2010). Challenges for nanoscale MOSFETs and emerging nanoelectronics. *Transsanctions Electronic Material*, *11*(3), 93–105. doi:10.4313/TEEM.2010.11.3.093

Knoch, J., & Appenzeller, J. (2008). Tunneling phenomena in carbon nanotube field-effect transistors. *Physica Status Solidi. A, Applications and Materials Science, 205*(4), 679–694. doi:10.1002/pssa.200723528

Knoch, J., & Appenzeller, J. (2008). Tunneling phenomena in carbon nanotube field-effect transistors. Physica Status Solidi, 205, 679-694. doi:10.1002/pssa.200723528

Koswatta, S. O., Nikonov, D. E., & Lundstrom, M. S. (2005). Computational study of carbon nanotube p–i–n Tunnel FET. *IEEE Internat Electron Devices Meeting, Technical Digest*, 518-521. 10.1109/IEDM.2005.1609396

Marani, R., & Perri, A. G. (2019). A Comparison of CNTFET and CMOS Technology through the Design of a SRAM Cell. *ECS Journal of Solid State Science and Technology, 8*(1), M1–M18. doi:10.1149/2.0141901jss

Nardelli, M., Fattebert, J.-L., Orlikowski, D., Roland, C., & Zhao, Q. (2000). Mechanical properties, defects, and electronic behavior of carbon nanotubes. Carbon, 38(11), 1703-1711.

Patel, P. K., Malik, M. M., & Gupta, T. K. (2019). Performance evaluation of single-ended disturb-free CNTFET-based multi-Vt SRAM. *Microelectronics Journal, 90*, 19–28. doi:10.1016/j.mejo.2019.05.015

Patel, P.K., Malik, M.M., & Gupta, T.K. (2018). Reliable high-yield CNTFET-based 9T SRAM operating near threshold. *Journal of Computational Electronics, 17*(2), 774–783.

Pourfath, M., Kosina, H., & Selberherr, S. (2007). Tunneling CNTFETs. *Journal of Computational Electronics, 6*(1-3), 243–246. doi:10.100710825-006-0099-1

Raj, B., Saxena, A. K., & Dasgupta, S. (2009). Analytical modeling for the estimation of leakage current and subthreshold swing factor of nanoscale double gate FinFET device". *Microelectronics International, 26*(1), 53–63. doi:10.1108/13565360910923188

Scarlet, S. P., Ambika, R., & Srinivasan, R. (2017). Effect of eccentricity on junction and junctionless based silicon nanowire and silicon nanotube FETs. *Superlattices and Microstructures, 107*, 178–188. doi:10.1016/j.spmi.2017.04.015

Seabaugh, A. C., & Zhang, Q. (2010). Low-voltage tunnel transistors for beyond CMOS logic. *Proceedings of the IEEE, 98*(12), 2095–2110. doi:10.1109/JPROC.2010.2070470

Shaker, A., Ossaimee, M., Zekry, A., & Abouelatta, M. (2015). Influenceof gate overlap engineering on ambipolar and high frequency characteristics of tunnel-CNTFET. *Superlattices and Microstructures*, *86*, 518–530. doi:10.1016/j.spmi.2015.08.008

Sharma, S. K., Raj, B., & Khosla, M. (2017). Subthreshold Performance of $In_{1-x}Ga_xAs$ Based Dual Metal with Gate Stack Cylindrical/Surrounding Gate Nanowire MOSFET for Low Power Analog Applications. *Journal of Nanoelectronics and Optoelectronics.*, *12*(2), 171. doi:10.1166/jno.2017.1961

Singh, A., Khosla, M., & Raj, B. (2017). Design and analysis of electrostatic doped Schottky barrier CNTFET based low power SRAM. *AEÜ. International Journal of Electronics and Communications*, *80*, 67–72. doi:10.1016/j.aeue.2017.06.030

Singh, J., Mohanty, S. P., & Pradhan, D. (2012). *Robust SRAM Designs and Analysis*. Springer Science & Business Media.

Singh, J., Ramakrishnan, K., Mookerjea, S., Datta, S., Vijaykrishnan, N., & Pradhan, D. (2010). A novel si-tunnel FET based SRAM design for ultra low-power 0.3V VDD applications. In *Proceedings of the 2010 Asia and South Pacific Design Automation Conference*. IEEE Press. 10.1109/ASPDAC.2010.5419897

Stanford University Nanoelectronics Group. (2018). *Stanford University CNFET model*. Retrieved from http://nano.stanford.edu/model. php?id=23

Sze, S. M. (2009). *Semiconductor Devices: Physics and Technology* (2nd ed.). Wiley India.

Tekleab, D. (2014, May). Device performance of silicon nanotube field effect transistors. *IEEE Electron Device Letters*, *35*(5), 506–508. doi:10.1109/LED.2014.2310175

Tekleab, D., Tran, H. H., & Sleight, J. W. (2012). *Silicon nanotube MOSFET*. U.S. Patent 2012217468 A1.

Yamauchi. (2008). Embedded SRAM Design and Trend. *IEEE ISSCC 2008, Memory forum Embedded Memory Design for Nano-scale VLSI system.*

Yamauchi, H. (2007). Embedded SRAM Circuit Design Technologies for a 45 nm and Beyond. *The 7th International Conference on ASIC (ASICON).*

Young, K. K. (1989). Short-channel effect in fully depleted SOI mosfets. *IEEE Transactions on Electron Devices*, *36*(2), 399–402. doi:10.1109/16.19942

Chapter 3
Carbon Nano Tube-Based Sensor Design for NEMS/MEMS Applications

Rekha Devi
Chandigarh University Mohali, I. K. Gujral Punjab Technical University, Jalandhar, India

Sandeep Singh Gill
National Institute of Teachers Training and Research, Chandigarh, India

ABSTRACT

This chapter deals with designing CNT-based piezoresistive pressure sensors with different boss sizes and with different configurations designed for low pressure range. The purpose for this work is to show a NEMS-based pressure sensor, which was analyzed by using ANSYS 17 software. The different combination of the diaphragm shows the improved performance of the pressure sensor in the case of CNT as compare to the silicon. This chapter is organized in sections, where section 2 discusses the review of CNT based MEMS/NEM design process and applications, Section 3 elaborates the use of CNT materials for design piezoresistive pressure sensors, Section 4 discusses mathematical modeling and simulation of CNT-based piezoresistive pressure sensors, Section 5 examines the results and discussion in terms of linearity and sensitivity of designed sensor, and Section 6 consummates the chapter with the conclusion.

DOI: 10.4018/978-1-7998-1393-4.ch003

1. INTRODUCTION

1.1MEMS & NEMS Technology

MEMS has been observed as best growing technologies for the current century and has the capability to revolutionize both consumer and industrial products by combining CNT and silicon-based technology. NEMS (Nano Electro Mechanical) have the structure elements less than or below 100nm, this distinguishes NEMS from MEMS (Micro Electro Mechanical). There are different types of NEMS based devices, which are as from the simple stationary structure to the highly complex Nano system that contain the numbers of moveable parts under the control of integrated nanoelectronics. . The mostly used material for the NEMS and MEMS is the CNTs and Silicon because of its electrical properties, mechanical properties, physical properties as well as electromechanical properties.

1.2 MEMS & NEMS Components

MEMS represent integrated systems of very small size where the feature sizes are reducing to micron and nano dimensions. MEMS have unique features of sensing, actuation, manipulation, control and computation. Which all are integrated in the same system more important than this "size" characterization. Nano sensor, Nano electronics, Nano actuator, Nano structure are the main components of MEMS are given below in figure 1.Recently, NEMS and MEMS developers and researchers have demonstrated large number of Nano sensors, nanoactuators, micro sensors and micro actuators with improved performance. The size of MEMS device is in the range of 20 micrometers to millimeter and size of its sub-components is in the range of 1 to 100 micrometers whereas the size of NEMS devices are at the molecular level having the structure element less than or below 100nm. Typical active substrate material used for actuators and for sensors in microsystems or other MEMS components include silicon, gallium arsenide, germanium and quartz whereas the NEMS components includes the carbon Nano tubes, carbon Nano ribbons, graphene, etc. These are also beneficial for sensing devices, which can be good for the devices.

The functional elements of NEMS are miniaturized sensors, actuators, mechanical structure along with the microelectronics as shown in figure 1, among them Nano sensors and Nano actuators are the most notable elements.

Figure 1. Schematic illustrations of MEMS & NEMS components

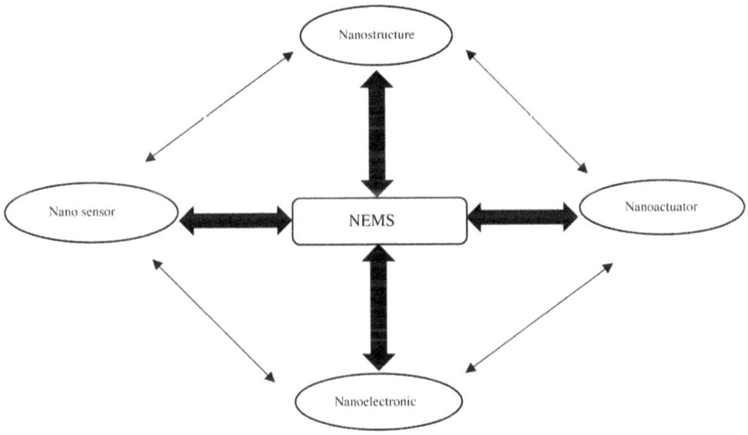

2. NANO SENSOR

Nano sensors are used to observe the intensity of certain biological, physical and chemical quantities in the environment of system. These Nano devices sense the fluctuation in their atmosphere and then pass the information for processing to the Nano electronics and in response produce an electrical signal as the input signal such as temperature, pressure, magnetic flux, chemical composition, sound and nuclear radiation (Hsu, 2008). Nowadays, the most commonly used Nano sensors are the thermal sensor, bio sensor, chemical sensor, pressure sensor and optical sensor (Tuantranont & Bright, 2000, p. 227-236). The pressure sensor is considered as the most notable and most interesting sensor (Yunus et al., 2015). The front part of the NEMS device is the sensing element. The diaphragm of silicon material with four piezo resistors of CNT material are good for stress and deformation detection.

2.1 Nanoelectronics

Nanoelectronics is the major processing unit of the device or of the system same as brain, which gets information and then process it accordingly. In the MEMS the information begins from the Nano sensors. The processed information is then passed by the process unit to Nano actuators which leads to some fluctuations in its atmosphere.

2.2 Nanoactuator

Nanoactuator is a device which converts an electrical signal to the mechanical signal especially the Nano scale movement (Yunus et al., 2015). The end part of the MEMS device is the Nano actuating that leads by moving, rotating, positioning, regulating, pumping and filtering, for controlling the atmosphere for some kind of purpose or results of device.

2.3 Nanostructure

MEMS devices and systems are very small in size. Motors, levers, gears, filters, diaphragms, steam engines and piston are the extremely small structures can be fabricated on the surface of a silicon chip. These Nano structures can be used as actuating and sensing elements in Nano actuators and Nano sensors respectively.

3. APPLICATIONS OF MEMS & NEMS

The table below discusses the general domains in which NEMS devices are used and the possibilities for MEMS technology in these areas.

Table 1. NEMS devices and the possibilities for MEMS technology in these areas

S. No	Area	Applications
1.	**Aerospace**	Micro-satellites, Command and control microelectronics systems, Tunable mirror arrays for adaptive optics, Propulsion and attitude control.
2.	**Automotive**	Air Bag Accelerometers Systems, Headlight Leveling, Inertia Brake Lights, Active Suspension control accelerometers, Rollover Detection, Air conditioning compressor sensor, Internal navigation sensors, and Car tire pressure sensors.
4.	**Communications**	RF relays, switches and filters, Fibers-optics network components, Voltage controlled oscillators, Splitters and couplers, Tunable lasers, Optical switching technology for data communication.
5.	**Commercial Electronic**	Inkjet printers, DMD chip in a projector, Game controllers, Personal media players, Interferometric modulator display for mobiles, Digital Cameras, Computer Peripherals, Mirror Arrays for Displays.
6.	**Defense**	Surveillance, arming systems, Aircraft control, Identification friend or foe, Munitions guidance, Nano robotic electronics disabling system, chemical attack warning sensor, Distributed battlefield sensor net.
7.	**Medical Electronics**	Blood pressure sensors, for DNA identification and amplification Polymerase Chain Reaction systems, Scanning Tunneling Microscopes, Pacemakers, Implanted pressure sensor communication device.

4. NEMS PRESSURE SENSOR

In the pressure sensor, the input to the sensor are feed through the pressure which act like an input is a necessary parameter in different real time applications. As, the pressure applied force or pressure per unit area is defined as pressure. Mathematical expressions are as P= F/A, Where A= The area on which force is applied F= Force applied. Pressure sensors also used for directly and indirectly measure other variable like fluid flow, water flow, gas flow, water level, speed and altitude.

Pressure sensors are commonly used as force collector such a bourdon tube, piston, bellows or diaphragm to measure deflection or strain with applied pressure and also use s density to sense the pressure of a gas or liquid . Approximately, the diaphragm is the best sensing element in all types of micro pressure sensor, diaphragm-based sensor have deflection in response of applied pressure but they attains very small mechanical deflection which cannot be measured directly. This deflection or strain generated in the diaphragm is transformed into electrical signals using an appropriate transduction mechanism namely piezoelectric, capacitive or piezo resistive. Thus, pressure sensor can also be classified on the basis of its transduction mechanism includes

1. **Capacitive Pressure Sensor:** A capacitance pressure transducer senses and measures the change in capacitance among a flexible metal diaphragm and a fixed metal electrode. These two electrodes are separated by spacer or dielectric material in parallel plate capacitive pressure sensor (Bhat & Nayak, 2013, P.1-10; Bao, 2005). Figure 2 shows a schematic diagram of a silicon micro machined capacitive pressure sensor.

The advantage of capacitive pressure sensors is their high sensitivity, low thermal sensitivity, low noise, low power consumption and large dynamic range as compared to other pressure sensor.

Figure 2. Schematic representation of capacitive pressure sensor

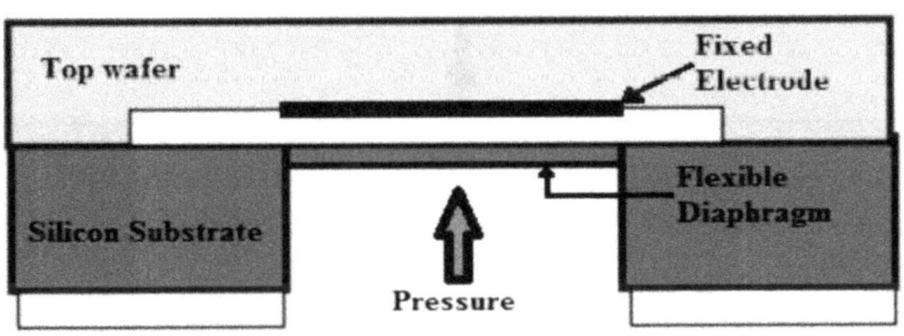

2. **Piezoelectric Pressure Sensors:** A piezoelectric pressure sensor causes change in its electrical potential proportional to the applied pressure (Barlian, 2009). This type of sensors are highly suitable for dynamic pressure sensing and are inappropriate for the measurement of static pressure as piezoelectric materials only respond to varying strains. The main advantage of this approach is that there is no requirement of an external excitation source such as power supply.
3. **Piezoresistive Pressure Sensors**: In this approach, much better sensitivity has been attained since the piezo resistors are embedded directly on the silicon diaphragm in the selected regions of maximum stress as shown in Figure 3. Basically, the piezoresistive pressure sensor consists of an elastic diaphragm and piezoresistive element. This sensor simply works on the phenomenon of change in electrical resistance in response to mechanical stress (Smith, 1994, p. 42)

As a pressure is applied, the diaphragm is deflected and deformed as well as the piezoresistive material. This means that applied pressure causes the piezoresistive material to become slightly thinner and longer as shown in Figure 4. Which cause the piezoresistive element to change its resistance and the mathematical expression of the resistance, R is given by:Where ρ, L, A, W, t corresponds to resistivity, length,

Figure 3. Schematic representation of piezoelectric pressure sensor

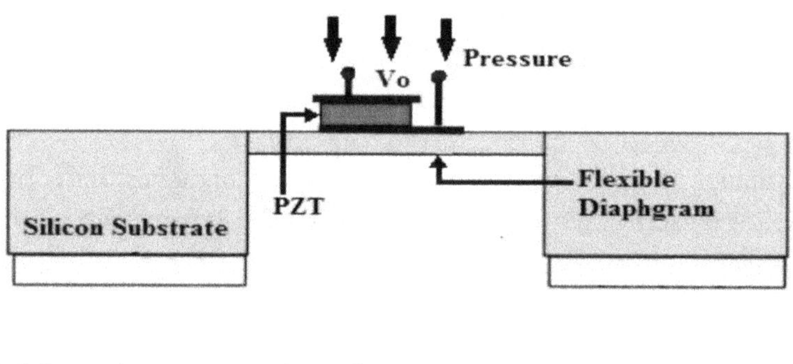

Figure 4. Schematic representation of piezoresistive pressure sensor

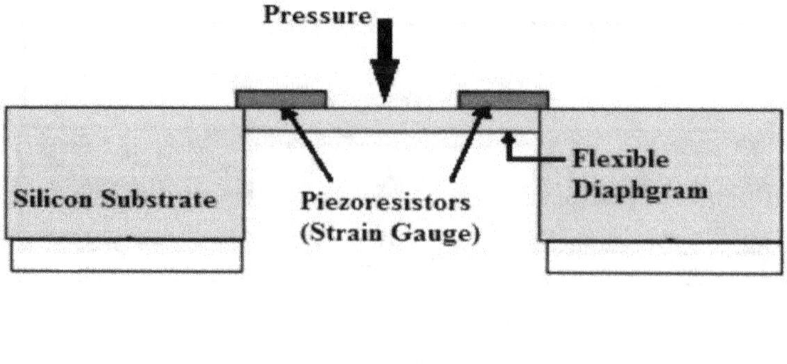

cross-sectional area, width and thickness of the piezoresistive material, respectively. When a pressure is applied, the piezoresistive material changes its dimensions and resistivity of material, which results into change in its resistance (Kumar & Pant 2015, p. 1949). Since the piezoresistive material becomes thinner and longer under the influence of pressure, from (1.2), it can be assumed that the resistance of the piezoresistive material is increased.

5. A REVIEW OF CNT BASED MEMS/NEMS DESIGN PROCESS AND APPLICATIONS

In the early 1990s, the discovery of CNTs has stimulated. CNTs hold outstanding potential applications in the industry of nanotechnology due to their special nearly one-dimensional structures, inherently unique chemical, electrical, exceptional mechanical and physical properties. on the performance of CNT-based nanodevices or Nano sensor, the buckling instability and vibration have a significant impact among the mechanical behaviors (Natsuki, 2017, p.56) CNTs do not break easily, they can buckle like straws. In fact, without any damage to the physical structure, they can easily be straightened. Young's modulus of materials has significant effect on its stiffness or flexibility under tensile strain. The Young's modulus of CNTs ranges from 1.28 - 1.8 TPa (tera pascals). (Sankar & Kumar, 2011, p.342) due to the C–C bond, CNTs are the stiffest and strongest material. Thermally, in both air and vacuum CNTs have high thermal stability. Depending upon the chirality (the direction of rolling the graphite sheet to form the carbon nanotube) and the tube diameters, CNTs can be categories either as semiconductor or metallic. An integer pairs (nm.) represent the chirality of carbon Nanotubes with $n - m = 3j$ (j being a nonzero integer) are metallic while all the others are semiconducting.

Carbon nanotubes (CNTs) have capacity to carry very high currents with almost negligible heating effect because of its highly anisotropic dielectric property. This only possible due to one-dimensional structures of nanotubes (Wang &Yeow, 2009).

It has been documented and investigated that for strain or pressure sensor CNTs uses as a novel sensing medium. The electrical properties and mechanical deformation of CNTs are used as sensing medium for measurable piezoresistive effect. As compared to doped polysilicon the CNTs are beneficial strain/pressure sensor because of its elasticity, higher gauge factor and bending to small radius without damage the tube.

5.1 Piezoresistance of CNTs

The piezoresistive sensing property of CNTs material is defined by the gauge factor (G_F), which is the ratio of change in resistance to the strain(Є).

$$G_F = \frac{\Delta R}{R}\frac{1}{\in}.$$

The strain-dependent band gap (E_g) of a ballistic SWNT is expressed as

$$E_g = E_g^0 + \frac{dE_g}{d\in} \in \cdot E_g^0 = \frac{2_Y a}{d\sqrt{3}} \cdot \frac{dE_g}{d\in} = sign\left(2p+1\right)3_Y[\left(1+v\right)cos3\theta 1$$

E_g is the initial bandgap energy. E_g^0 .s unstrained bandgap (~0.1 to 2 eV)for an average semiconducting CNTs, γ=2.6 eV, is the tight binding overlap integral, d is the diameter of the CNTs, and a is the unit vector of graphene. dEg /dε is the change in bandgap, v is the Poisson's ratio of the tube ~0.20 and θ1 is the chiral angle of tube and The sign of the equation depends on p(p= 0, ±1) labels the CNTsfamily. The familyindex pis given by the chiral index (*n, m*), where n − m = 3q + p, where q is an integer and p is −1, 0, or +1. In case of metallic CNTs, the bandgap belonging to the family index p = 0 rise with strain, while semiconducting CNTs of family index p = +1 also shows an rise in bandgap with strain. The semiconducting CNTs of family index p = −1 shows a decrease in band gap with strain. A maximum bandgap changes of dEg /dε~95 meV/%.

The resistance of a CNTs depends on its energy bandgap (E_g) and is given as

$$R_{tot}\left(\in\right) = R_C + \frac{1}{|t|^2}\frac{h}{8e^2}\left[1 + exp\left(\frac{E_g}{kT}\right)\right].$$

R_C is the contact resistance in series, |t| is the absolute transmission coefficient, h is Plank's constant, e is the electron's charge of the electricity, k is the Boltzmann constant, T is the absolute temperature. The gauge factor of the CNT sensor is finding with the resistance R_{tot} at zero strain (Selvin et al., 2011, p 3; Obitayo & Liu, 2012).

5.2 Placement of Piezo Resistors

The basic principle of carbon nanotube-based square piezoresistive pressure sensor is to measure the change in the resistance of the CNTs material due to the applied pressure. The application of pressure below the sensor engender a deflection of the diaphragm and this causes the change in resistance of the CNTs. (Patel et al., 2016) The pressure sensor design consists of four piezoresistive CNTS elements placed on top surface of silicon square diaphragm. The placement of these four CNTs piezo resistors should be in the high-stress regions (HSR) within the Squared diaphragm. In the case of thin squared diaphragm, the HSR will be at the center of all the edges of the square diaphragm. Consequently, the CNT spiezo resistors are placed at the center of each edges of the squared diaphragm. The FIGURE (a). shows the zero-turn configuration of CNTs piezoresistive squared pressure sensor. (b) one turn (c) two turn and (d) three turns. copper, aluminum material would be use to provides the connections between four piezoresistors because these materials does not contribute in piezoresistive effects.it is founded from the literature the two-turn configuration have higher sensitivity and zero turn have better linearity (Kumar et al., 2013, p.339; Mohamad & Ibrahim 2016, p.1).

5.3 Resistance-Pressure Relationships

With applied an even small pressure, there is large increase in charge carrier's concentration course to lead larger electrical conductivity and hence decrease in resistance change of the thinner sample. This decrease in percentage of resistance can be calculated by the given equation percentage of decrease in resistance

$$= \frac{R_i - R}{R_i} * 100$$

.where R_i is the initial resistance and R is the resistance at maximum pressure.

The sensor's resistance at maximum pressure (R) can be calculated as

$$R = \frac{d\acute{A}}{A} = \frac{d}{\tilde{A}A}$$

where d is the distance between interelectrode or thickness of the pellet, A is the cross-sectional of pellet, and ρ is the resistivity ($\rho = 1/\sigma$), σ is the conductivity of the pellet. The total resistivity change of the material due to both the geometric effects $(1 + 2\upsilon)$ and resistivity change $(\Delta\rho/\rho\varepsilon)$ can be expressed as

Figure 5. Piezoresistor's different configurations (a) zero turn, (b)one turn,(c) two turn,(d) three turn

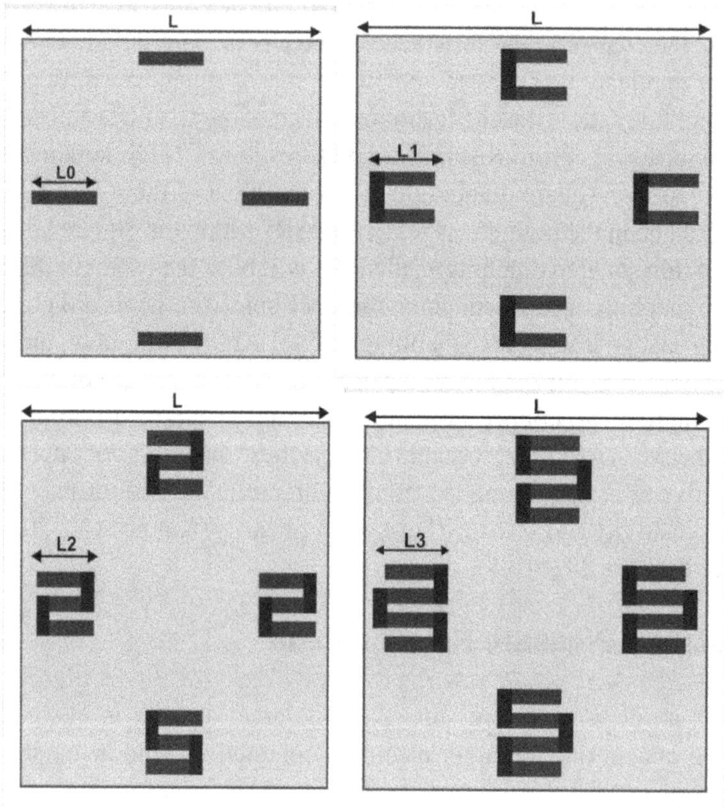

$$\frac{\Delta R}{R_0 \in} = \left(1 + 2v\right) + \frac{\Delta p}{p \in}.$$

$$\frac{\Delta R}{R_0} = \left(1 + 2v\right) \in + \frac{\Delta p}{p}.$$

In case of nanomaterials, the geometrical parameter is negligible because the change in resistivity have large enough as compared to the geometrical factors.(Ali et.al., 2018)

5.4 Methodology

CNTs undergo changes in their band structures when these are subjected to mechanical deformations. This phenomenon of deformation makes them applicable for pressure /strain sensing applications. The electrical resistivities of CNTs thin film increase linearly with strain, making it an ideal material for a piezoresistive Pressure/ strain sensor. The pressure sensor design consists of placing four piezoresistive CNT elements in the form of Wheatstone bridge on top of Silicon/Si3N4 diaphragm. thus, measuring the resistance of the CNTs. The applied pressure underneath the sensor causes a deflection of the silicon diaphragms and this causes a change in resistance of the CNTs. The optimal location to place the CNT would be the center of the edge's square diaphragms of maximum strain. As a result, the calculation of strain distribution and deflection in accordance with the applied pressure becomes pivotal.

5.5 IMPORTANT DESIGN PARAMETER OF PRESSURE SENSOR

1. **Stress Analysis (σ):** Stress is experienced in the diaphragm with the applied pressure and maximum stress produced at the center of the edges of the diaphragm, where the piezo resistors are placed. Change in resistance can be calculated as the function of the stress and piezoresistive coefficient. P is the applied pressure, υ. is the Poisson's ratio, $\pi l, \sigma x$ and $\pi t, \sigma y$ are the piezoresistive coefficient and stress respectively in longitudinal and in transverse direction (Patel et al., 2016). Pressure Sensor's s sensitivity is directly depending upon stress intensity in longitudinal and transverse directions (Kumar et al., 2013, p. 339). The resistance change should be maximized with the component of longitudinal stress and the resistance change kept as minimum caused by the component transverse stress (Mohamad, Soin, & Ibrahim, 2016, p.1).

$$\pi l \sigma_l + \pi t \sigma_t$$

$$\sigma_y = \upsilon \sigma_x$$

2. **Total Displacement (D):** Applied pressure on the diaphragms have tendency to deform the diaphragms and it is observed that the maximum deformation occurs at the center of the diaphragm and minimum at the fixed ends of the diaphragm. The deflection of diaphragm is inversely proportional to the thickness of the diaphragm. Total deflection at the center is calculate by the equation given below.

$$W = \frac{0.00126Pa^2}{D}$$

where D is flexure rigidity of the material, a is side length of the diaphragm and P is the uniform applied pressure. Where E is the equivalent Young's modulus and h is the diaphragm thickness and ν is the Poisson's ratio.

3. **Output Voltage (V_{OUT}):** The output voltage variation(ΔV) in the Wheatstone bridge circuit of the pressure sensor was due to applied stress. Mechanical stress caused the change in resistances (R1, R2, R3, and R4) of the four piezo resistors on the silicon diaphragms.

$$V_O = V_{in}\left[\frac{R_2R_3 - R_1R_4}{(R_1 + R_2)(R_2 + R_4)}\right] Volts\,.$$

Sensitivity(S): Sensitivity of the pressure Nanosensor is defined as the relative change in the output voltage of Wheatstone bridge circuit per unit of applied pressure. In case of pressure sensor, the sensitivity can be expressed as the minimum input pressure that will create a detectable change in output voltage. Pressure Nanosensor sensitivity S can be calculated by the equation given below (Patel et al., 2016):

$$S = \frac{(\text{"V})}{\text{"P}} * \left(\frac{1}{V_{in}}\right) = \frac{\Delta R}{\Delta P}\frac{I}{R} \quad ..$$

Non-linearity (NL): Nonlinearity is defined as the deviation of calibrated point from the end-point straight line. Thus, for every observation, there is a particular deviation and corresponding non-linearity. The non - linearity vs Diaphragm can be calculated as (Kumar et al., 2013, p. 339)

$$NL = \frac{V_o(P_i) - \frac{V_o(P_m)}{P_m}P_i}{V_o(P_i)} X100\%$$

6. DESIGN AND RESULTS

6.1 Design of Squared Piezoresistive NEMS Pressure Sensor With Diaphragm as Whole CNT

Piezoresistive transduction mechanism employs in the proposed pressure sensor where piezoresistive are supposed placed on the surface of the diaphragm at the maximum stress region. In this work applied pressure (0.01-0.11Mpa) and terminal voltage (5V) are the inputs. Deformation and stress are the output which are responsible for the sensitivity and linearity of the pressure sensor. The base model of square diaphragm is considered as a flat thin diaphragm of 10 μm thickness and 2000 μm edge length. In this, we have analyzed two case as whole CNT materials and only piezo resistors of CNT. Isotropic properties of silicon, polysilicon, and CNT are chosen. The design simulations for the clamped diaphragm are carried out by ANSYS 17.

The CNT material is assigned to the whole plate including the boss, the fixed part and the membrane.

The graphical representation shows the maximum deformation in the without boss structure whereas the others are having the same values approximately. As in the graphical representation, the boss structure values are clash behind each other's.

The graphical representation shows the maximum stress in the boss structure with the boss size of 1,50,000 nm. The minimum stress shows in the boss structure with the boss size of 50,000 nm.

Figure 6. Simulated results of whole diaphragm with boss size of 50,000 nm by applying 0.1 MPa pressure (a) Deformation (b) stress

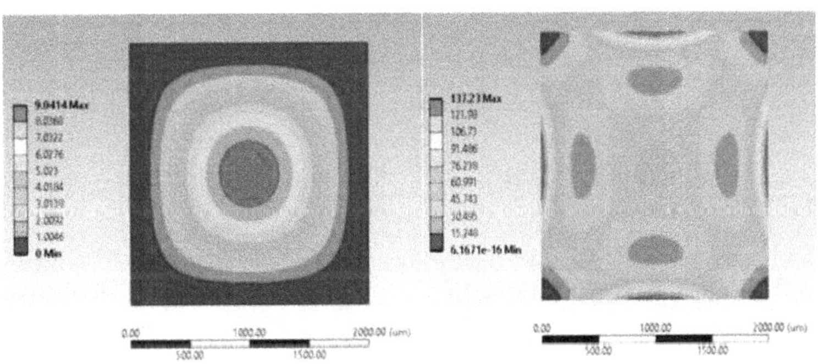

(b)

(a)

Figure 7. Simulated results of whole diaphragm without boss by applying 0.1 MPa pressure (a) Deformation (b) stress

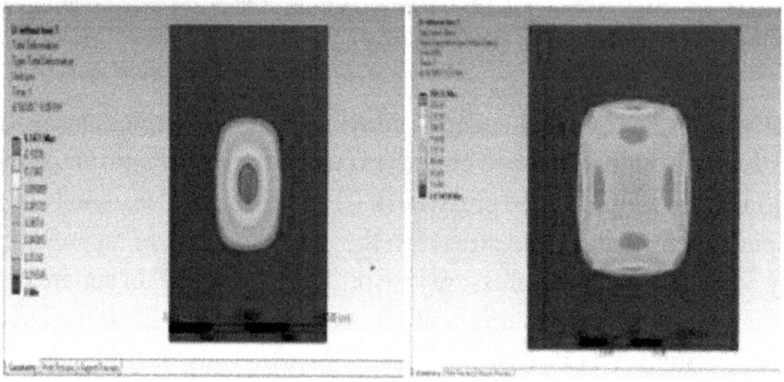

Table 2. Results for deformation (nm) of diaphragm for applied pressure range

Pressure	Without Boss	With Boss of 50,000	With Boss of 10^5	With Boss of 1,50,000
0.01	14.7	8.19	8.3	8.27
0.02	29.4	16.3	16.6	16.55
0.03	44.1	24.59	24.9	24.8
0.04	58.8	32.79	33.2	33.11
0.05	73.5	40.99	41.5	41.39
0.06	88.2	49.19	49.8	49.6
0.07	102.9	57.39	58.1	57.95
0.08	117.6	65.5	66.4	66.23
0.09	132.3	73.7	74.7	74.5
0.1	147.1	81.9	83	82.7
0.11	161.8	90.19	91.3	91

SUMMARY

The Squared diaphragms designed with thickness and edge length of 10µm and 2000 µm to measures the pressure in the range of(0.01 -0.11) MPa. The CNT material is used for the better results and the better performances of the device or the system with the Nanotechnology for the low-pressure range applications. The different cases are discussed with combination of material such as CNT and Silicon. By applying the different pressures on the diaphragm, the change in voltage will be calculated by varying the pressure. In the structure of diaphragm, the different boss size is

Figure 8. Graphical representation of deformation of diaphragm

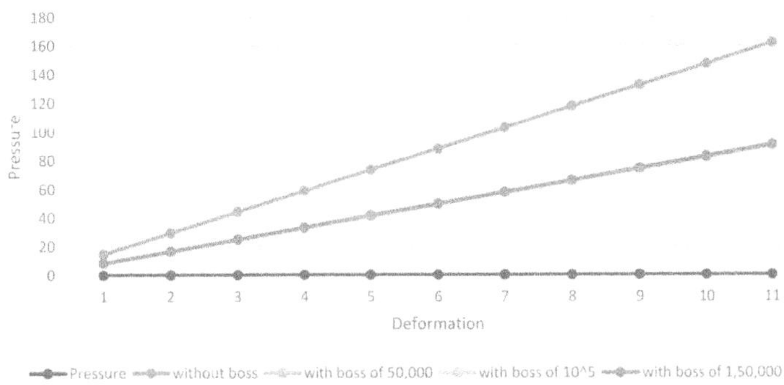

Table 3. Results for stress (nm) of diaphragm for applied pressure range

Pressure	Without Boss	With Boss of 50,000	With Boss of 10^5	With Boss of 1,50,000
0.01	28578	22543	28540	29316
0.02	57164	45083	57077	58628
0.03	85759	67619	85611	87938
0.04	114360	90150	114140	117240
0.05	142970	112680	142670	146550
0.06	171590	135200	171190	175850
0.07	200220	157720	199710	205140
0.08	228860	180240	228230	234440
0.09	257500	202750	256740	263730
0.1	286160	225260	285250	293010
0.11	314820	247760	313760	322290

Figure 9. Graphical representation of stress of diaphragm

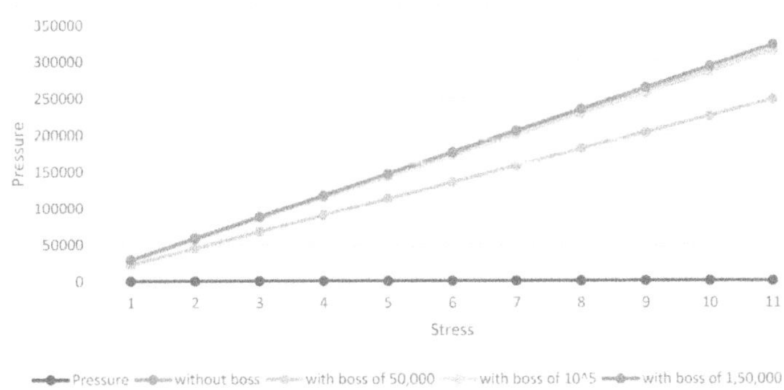

used to compare the different results on the square diaphragm. This proposed work will enhance the sensitivity and the non-linearity of the device as the sensitivity is directly proportional to the stress generated whereas the non-linearity error is directly proportional to the deformation which further depends upon the thickness of the membrane.

REFERENCES

Ali, A., Khan, A., Karimov, K. S., Ali, A., & Daud Khan, A. (2018). Pressure sensitive sensors based on carbon nanotubes, graphene, and its composites. *Journal of Nanomaterials*.

Bao, M. (2005). *Analysis and design principles of MEMS devices*. Elsevier.

Barlian, A. A., Park, W. T., Mallon Jr, J. R., Rastegar, A. J., & Pruitt, B. L. (2009). Review: Semiconductor Piezoresistance for Microsystems 2009. *Proc. of the IEEE, 97*(3).

Bhat, K. N., & Nayak, M. M. (2013). MEMS pressure sensors-an overview of challenges in technology and packaging. *Journal of Smart Structures and Systems, 2*, 1-10.

Hsu, T. R. (2008). *MEMS and microsystems: design, manufacture, and nanoscale engineering*. John Wiley & Sons.

Kumar, S. S., Ojha, A. K., Nambisan, R., Sharma, A. K., & Pant, B. D. (2013). Design and simulation of MEMS silicon piezoresistive pressure sensor for barometric applications. *Proceedings of the ARTCom&ARTEEPEIE&itSIP and PCIE*, 339-345.

Kumar, S. S., & Pant, B. D. (2015). Polysilicon thin film piezoresistive pressure microsensor: Design, fabrication and characterization. *Microsystem Technologies, 21*(9), 1949–1958. doi:10.100700542-014-2318-1

Mohamad, M., Soin, N., & Ibrahim, F. (2016, December). Design of a high sensitivity MEMS piezoresistive intracranial pressure sensor using three turns meander shaped piezoresistors. In *2016 International conference on bio-engineering for smart technologies (BioSMART)* (pp. 1-4). IEEE. 10.1109/BIOSMART.2016.7835596

Natsuki, T. (2017). Carbon nanotube-based nanomechanical sensor: Theoretical analysis of mechanical and vibrational properties. *Electronics (Basel), 6*(3), 56. doi:10.3390/electronics6030056

Obitayo, W., & Liu, T. (2012). A review: Carbon nanotube-based piezoresistive strain sensors. *Journal of Sensors*.

Patel, R. S., Ganapathiraj, S., Kundaragi, S., & Gadekar, P. (2016). Carbon Nanotubes Based Piezoresistive Pressure Sensor. *Proceedings of 22nd IRF International Conference*.

Sankar, P. G., & Kumar, K. U. (2011). Mechanical and electrical properties of single walled carbon nanotubes: A computational study. *European Journal of Scientific Research*, *60*(3), 342–358.

Selvin, S. A., Lovelin, S. A., Moorthy, N. B., Gupta, A., Alagappan, M., & Ramalingam, V. (2011). Design and Simulation of Carbon Nanotube based Piezoresistive Pressure sensor. In *Expert from the proceedings of the 2011 COMSOL Conference in Bangalore, Page* (No. 3). Academic Press.

Smith, C. S. (1954). Piezoresistance effect in germanium and silicon. *Physical Review*, *94*(1), 42–49. doi:10.1103/PhysRev.94.42

Tuantranont, A., & Bright, V. M. (2000). Introduction to Micro-Electro-Mechanical Systems (MEMS) with Emphasis on Optical Applications. *NECTEC Technical Journal*, *1*(6), 227–236.

Wang, Y., & Yeow, J. T. (2009). A review of carbon nanotubes-based gas sensors. *Journal of Sensors*, 2009.

Yunus, N. A. M., Halin, I. A., Sulaiman, N., Ismail, N. F., & Sheng, O. K. (2015). Valuation on MEMS Pressure Sensors and Device Applications. *International Journal of Electrical, Computer, Energetic, Electronic and Communication Engineering*, *9*(8).

Chapter 4
CNTFET for Logic Gates Design

Gurmohan Singh
Centre for Development of Advanced Computing, India

Manjit Kaur
Centre for Development of Advanced Computing, India

Yadwinder Kumar
https://orcid.org/0000-0003-2938-0329
Yadavindra College of Engineering, Punjabi University, India

ABSTRACT

The novel characteristics of CNTFET have eliminated many technological and fundamental hindrances being faced by CMOS transistors. CNTFET is emerging as prospective replacement for CMOS transistors in digital circuits and systems. This chapter introduces design of CNTFET-based basic logic gates. The basic logic gates analyzed are inverter, NAND, and NOR gates. The designed gates are evaluated in terms of delay, power consumption, and figure-of-merit power-delay-product (PDP). The standard H-SPICE CNTFET model of Stanford University has been used for all simulations. The impact of dielectric material variations on performance parameters of carbon nanotube field effect transistor based universal gates has been analyzed. Comparison between CMOS and CNTFET-based logic circuits is carried out for different dielectric material at 16 nm technology node.

1.1 INTRODUCTION

Carbon Nanotube Field Effect Transistor (CNTFET) make use of carbon nanotubes (CNTs) as channel material while a traditional MOSFET uses bulk silicon as a channel material (Appenzeller, Lin, Knoch, Chen, & Avouris, 2005). It was first

DOI: 10.4018/978-1-7998-1393-4.ch004

experimentally testified in 1998, since then there is a continuous development in CNTFETs. If standard bulk silicon thickness is 0.5mm then from that prospective only 5-10 nm of area on the top surface level is being used for transistor action. For *n*-type or *p*-type CNTFETs, highly doped source and drain terminals are formed, for *n*-type group 5 elements such as phosphorus, arsenic and for p-type group 3 elements such as boron, gallium are used. CNT act as a channel. It is hollow inside with gate-all-around structure. It is used to mitigate short channel effect. Instead of using bunch of CNTs, SWCNT can be used to achieve area efficiency and low power consumption device. Placing of carbon nanotubes inside the trenches between source and drain is a difficult process. CNT conductivity depends upon its chirality; it can be semiconducting, metallic or semi-metallic in nature. 1-D structure of CNT makes scattering rare, which give rise to ballistic transport. CNT can operate at elevated temperature (~250°C), very high current carrying capability (~ 1010 A/ cm^2), larger mean free path etc. (Hoenlein, Kreupl, Duesberg, Graham, Liebau, Seidel & Unger, 2003). With the use of CNTs in FET high performance, high impedance, faster switching speed can be achieved. CNTFET structure with CNTs as a channel between source and drain terminal is shown in Figure 1.

Due to these remarkable advantages offered by CNTFETs, lot of research has been going on modeling and making its way towards digital circuits and systems to compute its performance (Raychowdhury, Mukhopadhyay, & Roy, 2004). CNTFET is one of the best promising technologies to substitute existing silicon technology because of following reasons:

1. The working principle and device structure is almost same as basic MOSFET device.
2. CNTFETs have excellent current carrying capability.
3. CNTs are light in weight and work on the principle of ballistic transport.
4. CNTFETs have sharp sub-threshold slope.
5. Improved current density (CNTs has current density nearly three times higher than that of silicon device).
6. Electron mobility will be high due to non-equilibrium transport mechanism.
7. High drive current.
8. No degradation down to 9 nm channel length.

Basically digital system design is associated with the binary logic based on two valued logic ('0' or '1') in Boolean space (Murotiya, Gupta, & Pandit, 2014; Lin, Kim, & Lombardi, 2011). Basic gate inverter and universal gates such as NAND and NOR gate forms the basics of any digital circuitry. Hence the optimum design of universal gates in terms of low power, lower delay, and good figure-of-merit (FOM) is desired for overall performance of digital circuit /system. Nowadays, speed and

Figure 1. Carbon nanotube field effect transistor structure

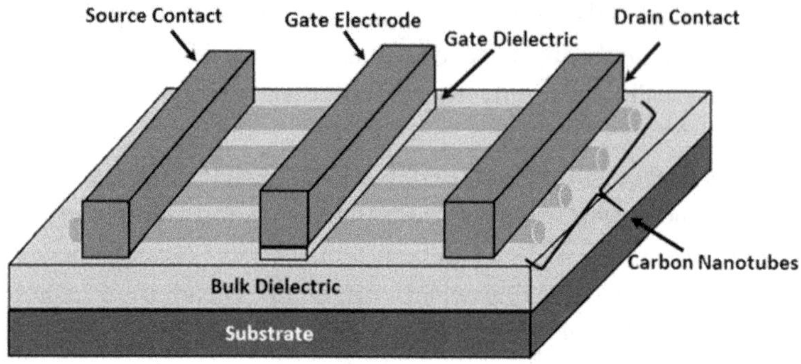

power are becoming most important issues (Qi, Kim, & Choi, 2012). CNTFET is emerging as a promising alternative to CMOS transistor to achieve low power and low off current properties (Appenzeller, 2008).

1.1.1 Types of CNTFETs

There are many different ways of placing CNT as a channel material in between highly doped source and drain terminals, with metal gate contact (Sinha & Chaudhury, 2015). The different types of CNTFET are:

1.1.2 Back Gated CNTFET

In this structure, metal strips used form the source and drain contacts separately. CNTs placed on two metal strips act as a channel between two contacts analogous to basic silicon MOSFET where silicon substrate act as a channel material. A metal contact is added at back of carbon nanotubes which make it gate able (Sinha & Chaudhury, 2015).

1.1.3 Top Gated CNTFET

In this structure, CNT is placed below the gate contact between source and drain regions. Number of top gated structures fabricated over a single wafer as gate contacts are isolated from one another as compared to back-gated CNTFET. Dielectric used at gate terminal is very thick (~15nm) which results in a very high electric field at very low gate voltage (Sinha & Chaudhury, 2015).

1.1.4 Gate-all-around CNTFET

In contrast to other structures where a part of CNT is connected to gate terminal, whole of the CNT is gated in gate-all around structure. CNT is wrapped inside a dielectric material with metal contacts at edges. It is also called as wrap-around gate CNTFETs, because almost entire part of the CNT is covered with metallic material (Sinha & Chaudhury, 2015).

1.1.5 Suspended Gate CNTFET

In this type of CNTFET, CNT is not attached to gate dielectric, it is suspend over trench. Dielectric at gate to be used is either air or vacuum. When gate voltage is applied, it attracts CNT towards gate and limits the space between gate and CNT creating a major drawback of suspended gate CNTFETs (Sinha & Chaudhury, 2015).

1.2 CNTFET SPICE MODELS

The pioneering work in developing a SPICE compatible model of CNTFET was initiated by Stanford University (Deng & Wong, 2007). They have developed various versions of CNTFET SPICE models. The following SPICE compatible model of CNTFET have been developed:

1.2.1 Stanford University Standard CNTFET Model

It is a SPICE-compatible compact model containing enhancement-mode, unipolar devices with semiconducting single-walled carbon nanotubes (SWCNT) as channels (Deng & Wong, 2007). The device in this model may have one or more CNTs with a specific user-defined chirality. The effects of scaling can be precisely modeled down to 20 nm. The model exhibit quasi-ballistic transport and covers an accurate depiction of the network of capacitors in a CNTFET. The model accurately incorporates many practical non-idealities, acoustic and optical phonon scattering, parasitic capacitance between source/drain and the gate, gate-to-contact-plug and gate-to-gate capacitances, charge screening effect, source/drain extension region resistance, Schottky-barrier resistance, and the band-to-band leakage current.

1.2.2 Stanford University Virtual Source CNTFET Model

It is a semi-empirical model which describes the current-voltage and charge-voltage (i.e. capacitance-voltage) characteristics in a short-channel MOSFET (Lee,

Franklin, Haensch, & Wong, 2015). In this model, the intrinsic drain current (Id) and terminal charges are computed assuming virtual source. The virtual source velocity for different channel lengths is extracted from experimental data. This model incorporates direct source-to-drain and band-to-band tunneling current calibrated by numerical simulations, parasitic capacitances including gate-to-contact coupling and gate-to-CNT fringe capacitances, and metal-to-CNT contact resistances calibrated by experimental data.

The major difference between VS-CNFET and the Stanford CNFET model is the modeling of carrier transport. The Stanford standard CNTFET model is physics-based having quasi-ballistic Landauer formula whereas VS-CNTFET model is based on the semi-empirical virtual source concept calibrated to experimental data .

1.2.3 Schottky Barriers CNTFET Model

This model examines the theoretical performance of CNTFETs using Schottky barriers (SBs) by means of a general ballistic model. It employs a unique approach to treat the SBs as mesoscopic scatterers by transforming the distribution functions for carriers in the channel. A significant reduction in current is observed compared to ballistic models without SBs. This model also incorporates quantum conductance degradation due to SBs.

1.3 OBJECTIVES OF THE CHAPTER

This chapter describes how CNTFETs take over MOSFET in terms of performance in digital circuits. Therefore, CNTFET need to be studied in detail and used as a best alternative to CMOS technology. Digital circuits like inverter, NAND and NOR gates are designed and simulated using H-Spice. Comparative analysis between circuits designed using standard CNTFET model and Predictive Technology Model (PTM) for CMOS transistors for low power applications at 16nm technology has been carried out. The effect of gate dielectric material and its thickness variations have been analyzed in terms of delay, power dissipation, and power-delay-product.

1.4 BACKGROUND

Lin, Kim, and Lombardi (2011) presented CNTFET based design of ternary logic gates by using binary gates which improved the performance, reduced area and power. (Shreya & Chandel, 2014) observed that CNTFET based digital circuits provide high speed, lower power consumption and are more energy efficient. (Yeole

& Padole, 2015) designed CNTFET and CMOS based logic gates and concluded that CNTFET based designs offer better figure-of-merit. (Sahoo, Sahoo, & Sankisa, 2015) presented CNTFET with optimum number of CNTs in channel and analyzed that in logic gates design SWCNT perform better than MWCNTs. Jamaa, Mohanram, and Micheli (2010) introduced a new technique to compute power consumption in ambipolar CNTFETs and demonstrated 57% power saving as compared to conventional CMOS. The authors proposed an ambipolar library of CNTFETs and concluded that it improve performance by seven times as compared to CMOS (Jamaa, Mohanram, & Micheli, 2011). Prabhu and Sarwade (2013) computed the effect of supply voltage variations on power-delay-product of CNTFET based logic gates. Kaur, Singh, and Kaur (2016) introduced virtual source CNTFET based basic logic gates and analyzed the effect of dielectric variations on performance parameters. The performance is computed in terms of delay, power consumption, and power-delay-product. Navin and Basavaraj (2013) presented CNTFET based memory cell and analyzed its performance. They further investigated that by reducing oxide thickness, metal contact size, and high-k dielectric improve switching speed of device. Singh, Khosla, and Raj (2016) investigated performance of CNTFET devices for achieving high I_{ON}/I_{OFF} ratio, low power applications, high transconductance, and sharp sub-threshold slope. Sinha and Chaudhury (2014) performed comparison of leakage power in CNTFET and MOSFET. Sethi, Kaur, and Singh (2017) introduced CNFET-based ternary content-addressable memory (TCAM) cell design and computed its performance in terms of PDP and static noise margin (SNM). The influence of chiral vector variations on performance of TCAM cell is also investigated.

1.5 LOGIC GATES DESIGN

This section illustrates schematic designs of CNTFET based basic logic circuits such as inverter, 2-input NAND, and 2-input NOR gates. These schematic designs have been used to verify the functionality in H-Spice simulator.

1.5.1 CNTFET Inverter

Figure 2 presents schematic design of CNTFET based inverter. The source and substrate of *p* type CNTFET are connected to power supply (VDD) and of n-type CNTFET are connected to ground (GND). The gate terminal of p-type and n-type MOSFET is given same input *In*. The output *Out* of circuit is drawn from common drain terminals of both p-type and n-type CNTFETs.

Figure 2. Schematic of CNTFET based inverter

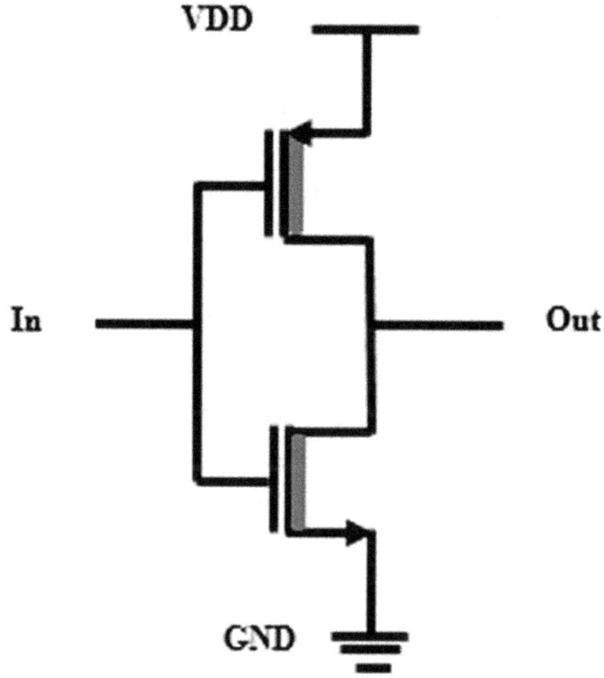

1.5.2 CNTFET NAND Gate

The schematic of CNTFET based 2-input NAND gate is shown in Figure 3. Two p-type CNTFET transistors connected in parallel to each other form the Pull Up Network (PUN). Similarly, two n-type CNTFET transistors connected in series form the Pull Down Network (PDN). The source and substrate of both PUN transistors are connected to VDD. Two inputs *In1* and *In2* are applied at inputs of circuits while *Out* is taken from common point of PUN and PDN. It is zero sensitive, when either of the input *In1* or *In2* is low (set to zero) then output is high, and output is low when both the input are set to high. Being a universal gate, any logic function can be implemented with NAND gates. The series connection of n-type transistors in PDN increase the resistance, so to make the transfer characteristics of 2-input NAND gate symmetric, the number of CNTs in PDN's n-type transistors have been doubled.

1.5.3 CNTFET NOR Gate

The schematic of CNTFET based 2-input NOR gate is shown in Figure 4. Two p-type CNTFET transistors connected in series form the Pull Up Network (PUN). Similarly, two n-type CNTFET transistors connected in parallel form the Pull Down

Figure 3. Schematic of CNTFET based 2-input NAND gate

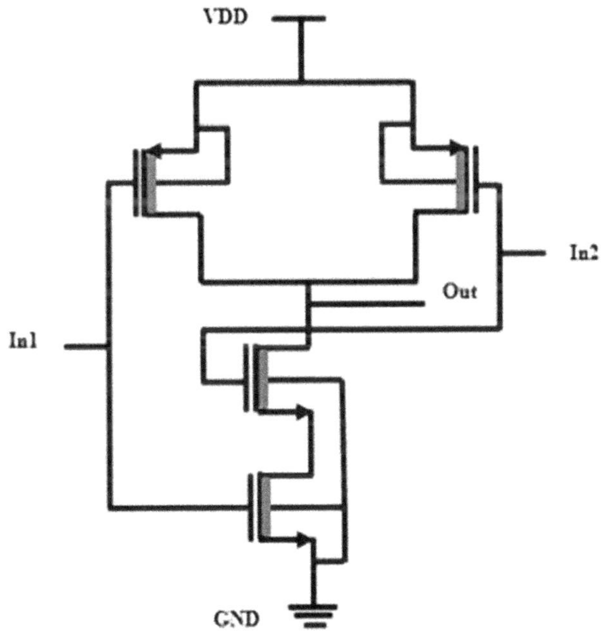

Network (PDN). The source and substrate of both PDN transistors are connected to GND. Two inputs *In1* and *In2* are applied at inputs of circuits while *Out* is taken from common point of PUN and PDN. It is one sensitive, when either of the input *In1* or *In2* is high (set to one) then output is low, and output is low when both the input are set to high. Being a universal gate, any logic function can be implemented with NOR gate. The series connection of p-type transistors in PUN increase the resistance, so to make the transfer characteristics of 2-input NOR gate symmetric, the number of CNTs in PUN's p-type transistors have been doubled.

1.6 SIMULATION RESULTS

All the simulations for CNTFET based logic gates i.e. inverter, 2-input NAND gate, and 2-input NOR gate have been performed using Stanford University Standard SPICE compatible model. The CMOS transistors model parameters are adopted from Predictive Technology Models (PTM). The simulation results are computed for CNTFET and CMOS based circuits at 16 nm technology node. The CNTFET based inverter, NAND, NOR gate characteristics have been plotted. The various performance parameters such as delay, average power consumption, and power-delay-

Figure 4. Schematic of CNTFET based 2-input NOR gate

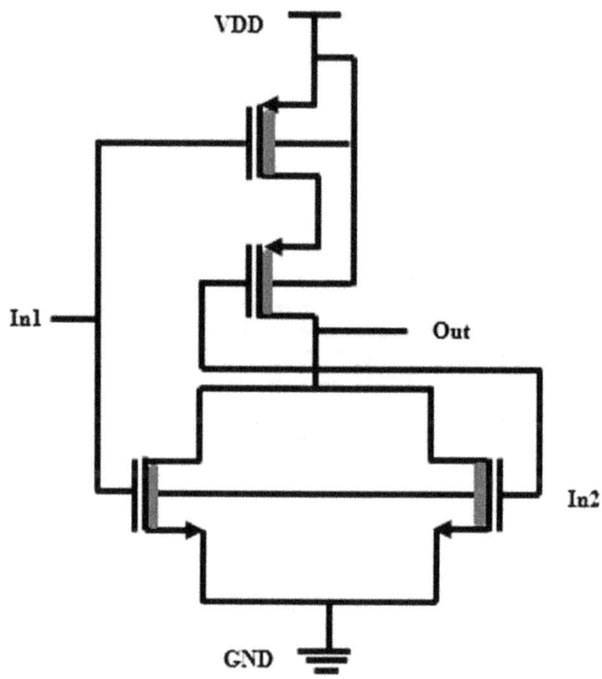

product are computed from simulation results. Also, the performance parameters are computed for different gate dielectric materials and by varying the thickness of the gate dielectric. The supply voltage of 0.9 has been used for all simulations. All transient simulations are performed by applying a transient input of frequency 1 GHz. All simulations are performed at room temperature of *25°C*. Inputs can be either Zero (Low) or High (One), thus output will be obtained as per logic operation of each gate.

The Table 1 lists parameter values of the CNTFET devices adopted for all simulations in this chapter.

1.6.1 Inverter Characteristics

The Figure 5 depicts transfer characteristics of inverter realization using CNTFET. The input voltage is varied from 0 to 0.9 volts. The transfer characteristics of inverter are highly symmetric and near to ideal characteristics. The switching point is about 0.45 V which is exactly midway of the output curve. The Figure 6 depicts transient characteristics of CNTFET inverter.

Table 1. CNTFET device parameter values used for all simulations

Device Parameter	Description	Value
Lch	Physical channel length	16.0 nm
Lss/Ldd	The length of doped CNT source/drain-side extension region	16.0 nm
Efi	The Fermi level of the doped S/D tube	0.6 eV
Kgate	The dielectric constant of high-k top gate dielectric material (planer gate)	3.9 to16.0
Tox	The thickness of high-k top gate dielectric material	1 to 4.0 nm
Csub	The coupling capacitance between the channel region and the substrate (backgate effect)	20.0 pF/m
Pitch	The distance between the centers of two adjacent CNTs within the same device	20.0 nm
(n1, n2)	The chirality of tube	(19, 0)

Figure 5. Transfer characteristics of CNTFET based inverter

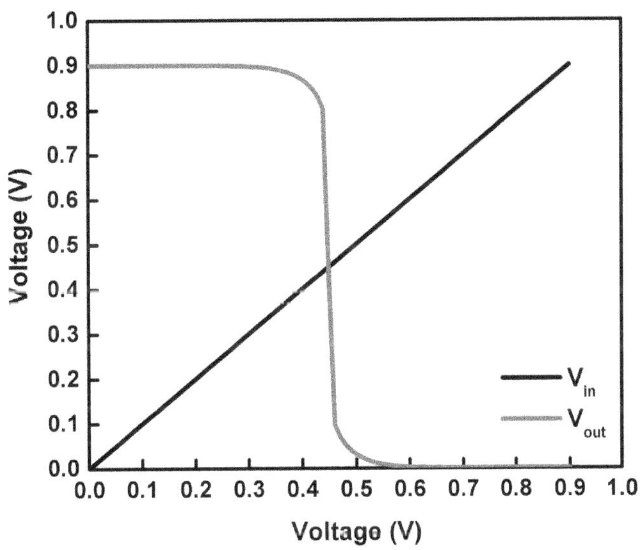

1.6.2 NAND Gate Characteristics

The NAND gate performs logic multiplication. If *In1* and *In2* are two inputs, then output *Out* is computed as $Out = \overline{In1.In2}$, where dot (.) represent AND operation and bar ($\bar{\square}$) represent invert operation. Output *Out* will be *High* when *Low* input is given at either of its input node or at both nodes and output will be *Low* for both

Figure 6. Transient characteristics of CNTFET based inverter

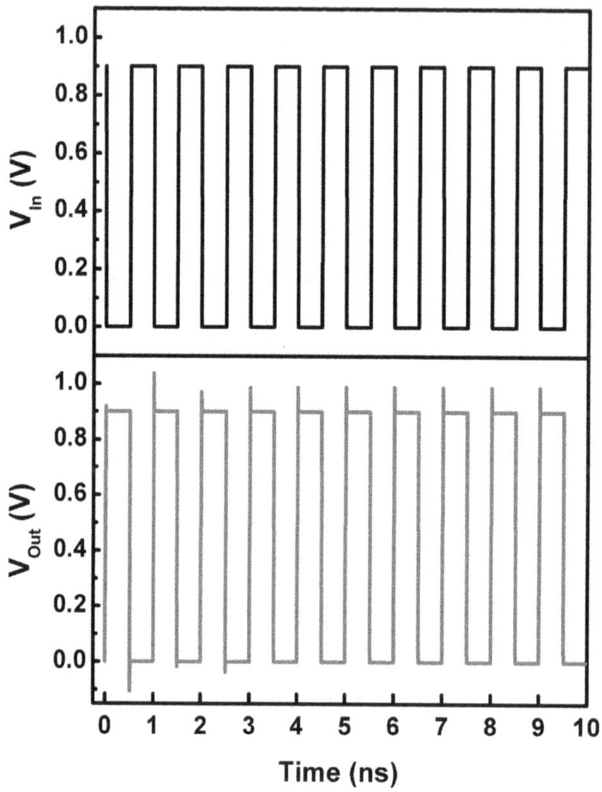

inputs at *High* level. It is basically a ZERO sensitive device i.e. output will be *High* for *Low* input. The Figure 7 depicts transient characteristics of CNTFET based 2-input NAND gate.

1.6.3 NOR Gate Characteristics

The NOR gate performs logical addition. If *In1* and *In2* are two inputs, then output *Out* is computed as $Out = \overline{In1 + In2}$, where dot (+) represent *OR* operation and bar ($\overline{\ }$) represent invert operation. Output *Out* is *High* only when both the inputs are *Low* otherwise output will be *Low*. It is basically a *High* sensitive device i.e. when either of its input is *High*, output will be *Low*. The Figure 8 depicts transient characteristics of CNTFET based 2-input NOR gate.

Figure 7. Transient characteristics of CNTFET based 2-input NAND gate

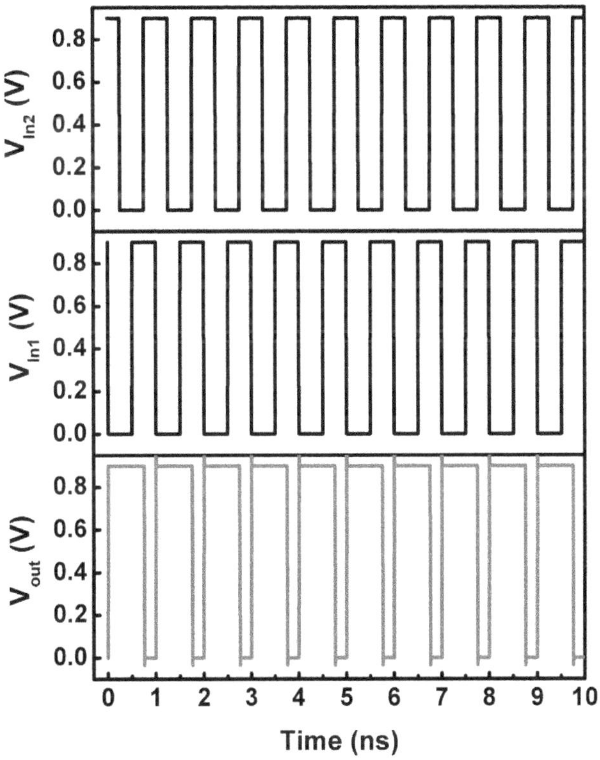

1.6.4 Performance Parameter Computations

The performance of CNTFET and CMOS based inverter, NAND, and NOR logic circuits have been computed for different gate dielectric materials. The dielectric materials such as Hafnium oxide (HfO_2), silicon nitride (Si_3N_4), and silicon dioxide (SiO_2) having dielectric constants 16, 7.5, and 3.9 are considered, respectively. Table 2 lists the delay, power dissipation and power-delay-product for CMOS and CNTFET based inverter circuits. The results reveal that CNTFET based inverter has smaller PDP value by a factor of approximately by 1000 times as compared to CMOS inverter. Similar, average power consumption of CNTFET based inverter is significantly small as compared to CMOS counterpart. The delay reductions for high-k HfO_2 and Si_3N_4 dielectric materials are 52.70 and 40.12%, respectively for CNTFET inverter as compareD to CMOS inverter. For SiO_2 dielectric material, delay degradation of 43.31% is observed for CNTFET as compared to CMOS inverter.

Figure 8. Transient characteristics of CNTFET based 2-input NOR gate

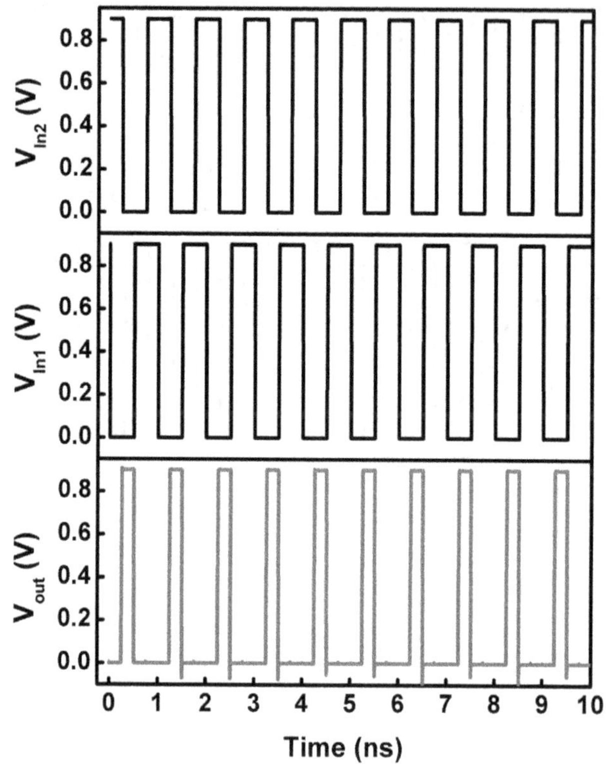

Table 2. PDP calculations for CNFET and CMOS based inverter at 16 nm technology

Dielectric Material	CMOS			CNTFET		
	Delay (ps)	Average Power (μW)	Power Delay Product (aJ)	Delay (ps)	Average Power (nW)	Power Delay Product (zJ)
HfO_2	2.275	35.162	80.015	1.076	6.729	7.246
Si_3N_4	1.398	35.041	49.014	0.837	2.602	2.179
SiO_2	1.018	34.981	35.615	1.796	2.125	3.818

Similarly, delay, average power consumption, and power-delay product for CMOS and CNTFET based NAND and NOR gates have been computed for different dielectric materials and listed in Table 3 and Table 4, respectively.

The dielectric thickness plays very important role in nanoscaled technologies to decide gate leakage currents. The non-ideal effects severely degrade performance

Table 3. PDP calculations for CNTFET and CMOS based NAND gate at 16 nm technology

Dielectric Material	CMOS			CNTFET		
	Delay (ps)	Power Dissipation (μW)	Power Delay Product (aJ)	Delay (ns)	Power Dissipation (nW)	Power Delay Product (aJ)
HfO_2	4.243	26.999	114.57	0.125	15.37	1.934
Si_3N_4	2.492	26.822	66.859	0.1261	8.186	1.032
SiO_2	1.812	26.732	48.446	0.127	5.807	**0.739**

Table 4. PDP calculations for CNTFET and CMOS based NOR gate at 16 nm technology

Dielectric Material	CMOS			CNFET		
	Delay (ps)	Power Dissipation (nW)	Power Delay Product (aJ)	Delay (ns)	Power Dissipation (nW)	Power Delay Product (aJ)
HfO_2	3.667	22.736	83.383	0.1257	14.850	1.867
Si_3N_4	2.674	22.254	59.521	0.1263	6.199	**0.7831**
SiO_2	2.135	22.065	47.125	0.1270	4.814	**0.6116**

of conventional CMOS transistors as dielectric thickness goes below 1.5 nm. Table 5, 6 and 7 illustrate PDP results of CNTFET based inverter, NAND and NOR gates for different dielectric materials as dielectric thickness is varied from 1 nm to 4 nm. The results reveal that as the dielectric constant increases, PDP at a fixed gate oxide thickness increases for high-k dielectric materials. It is also observed from PDP point of view Si_3N_4 offers best result.

1.7 FUTURE TRENDS

Since the first discovery and fabrication in 1991, CNT has emerged as most promising and attractive material for future nanoscaled technologies because of its physical, electrical and mechanical properties i.e. high current carrying capability, long ballistic transport, thermal stability and tensile strength. More fabricated devices data are required for statistical analysis. Also lot of work on reducing CNT-metal contact, gate dielectric material and its thickness need to carried out. Despite so much progress at modeling of CNTFET devices and their strong potential to

Table 5. PDP results for CNFET inverter for oxide thickness variations

Power Delay Product (zJ)			
Thickness of Oxide (T_{OX})	Dielectric Material		
	HfO_2	Si_3N_4	SiO_2
1 nm	7.732	3.912	2.268
2 nm	7.246	2.179	3.825
3 nm	3.621	2.949	3.641
4 nm	4.285	2.379	4.145

Table 6. PDP results for CNTFET based NAND gate for oxide thickness variations

Power Delay Product (aJ)			
Thickness of Oxide (T_{OX})	Dielectric Material		
	HfO_2	Si_3N_4	SiO_2
1 nm	2.740	1.453	0.825
2 nm	1.934	1.032	0.739
3 nm	1.413	0.555	0.763
4 nm	1.559	0.838	0.770

Table 7. PDP results for CNTFET based NOR gate for oxide thickness variations

Power Delay Product (aJ)			
Thickness of Oxide (T_{OX})	Dielectric Material		
	HfO_2	Si_3N_4	SiO_2
1 nm	2.350	1.191	0.7058
2 nm	1.867	0.7831	0.6116
3 nm	1.443	0.7417	0.6535
4 nm	1.445	0.6872	0.7683

replace CMOS transistors in future, still a well-organized framework and reliable investigations are deficient. The major directions could be to fabricate high reliability CNTFET in bulk with smaller per unit cost as compared to CMOS devices. Also, the performance improvements could be carried out by developing accurate device models and at through process level modifications.

1.8 CONCLUSION

This chapter presented performance analysis of CNTFET based basic logic gates such as inverter, NAND, and NOR gate circuits and compared them against CMOS counterparts. The performance is computed in terms of delay, average power, and power-delay-product using extensive HSPICE simulations. The standard CNTFET model from Stanford University has been employed for all simulations and PTM models have been used for CMOS transistors at 16 nm technology node. Also, the performance of CMOS and CNFET based inverter and NAND/NOR logic circuits have been investigated in detail for different gate dielectric materials such as Hafnium oxide (HfO_2), Silicon Nitride (Si_3N_4), and Silicon Oxide (SiO_2). The results reveal that CNTFET based inverter has smaller PDP value by a factor of approximately by 1000 times as compared to CMOS inverter. Similar, average power consumption of CNTFET based inverter is significantly small as compared to CMOS counterpart. The optimum results for PDP are obtained at gate oxide thickness of 2 to 3 nm. From results, it can be concluded that CNTFET devices have potential to replace CMOS devices to meet lower PDP requirements.

REFERENCES

Appenzeller, J. (2008). Carbon nanotubes for high-performance electronics—Progress and prospect. *Proceedings of the IEEE*, *96*(2), 201–211. doi:10.1109/JPROC.2007.911051

Appenzeller, J., Lin, Y. M., Knoch, J., Chen, Z., & Avouris, P. (2005). Comparing carbon nanotube transistors – the ideal choice: A novel tunneling device design. *IEEE Transactions on Electron Devices*, *52*(12), 2568–2576. doi:10.1109/TED.2005.859654

Ben Jamaa, M. H., Mohanram, K., & De Micheli, G. (2011). An efficient gate library for ambipolar CNTFET logic. *IEEE Transactions on Computer-Aided Design of Integrated Circuits and Systems*, *30*(2), 242–255. doi:10.1109/TCAD.2010.2085250

Deng, J., & Wong, H. S. P. (2007). A compact SPICE model for carbon-nanotube field-effect transistors including nonidealities and its application - Part I: Model of the intrinsic channel region. *IEEE Transactions on Electron Devices*, *54*(12), 3186–3194. doi:10.1109/TED.2007.909030

Deng, J., & Wong, H. S. P. (2007). A compact SPICE model for carbon-nanotube field-effect transistors including nonidealities and its application- Part II: Full device model and circuit performance benchmarking. *IEEE Transactions on Electron Devices*, *54*(12), 3195–3205. doi:10.1109/TED.2007.909043

Hazeghi, A., Krishnamohan, T., & Wong, H.-S. P. (2007). Schottky-barrier carbon nanotube field effect transistor modeling. *IEEE Transactions on Electron Devices*, *54*(3), 439–445. doi:10.1109/TED.2006.890384

Hoenlein, W., Kreupl, F., Duesberg, G. S., Graham, A. P., Liebau, M., Seidel, R., & Unger, E. (2003). Carbon nanotubes for microelectronics: Status and future prospects. *Materials Science and Engineering C*, *23*(6-8), 663–669. doi:10.1016/j.msec.2003.09.153

Jamaa, M. H. B., Mohanram, K., & Micheli, G. d. (2010). Power consumption of logic circuits in ambipolar carbon nanotube technology. In *Proceedings of Conference on Design, Automation & Test in Europe Conference & Exhibition* (pp. 303-306). IEEE.

Kaur, R., Singh, G., & Kaur, M. (2016). Effect of dielectric variations on performance of carbon nanotube field effect transistor based basic logic gates. *IOSR Journal of VLSI and Signal Processing*, *6*(4), 78–84. doi:10.9790/4200-0604027884

Lee, C. S., Pop, E., Franklin, V. W., Haensch, W., & Wong, H.-S. P. (2015). A compact virtual source model for carbon nanotube FETs in the sub-10-nm regime—Part I: Intrinsic element. *IEEE Transactions on Electron Devices*, *62*(9), 3061–3069. doi:10.1109/TED.2015.2457453

Lin, S., Kim, Y. B., & Lombardi, F. (2011). CNTFET-based design of ternary logic gates and arithmetic circuits. *IEEE Transactions on Nanotechnology*, *10*(2), 217–225. doi:10.1109/TNANO.2009.2036845

Murotiya, S. L., Gupta, A., & Pandit, A. (2014). CNTFET-based low power design of 4-input ternary XOR function. In *Proceedings of international conference on computer and communication technology* (pp. 347-350). IEEE.

Murotiya, S. L., Gupta, A., & Pandit, A. (2014). CNTFET-based low power design of 4-input ternary XOR function. *Proceedings of IEEE International Conference on Computer and Communication Technology (ICCCT)*, 347-350. 10.1109/ICCCT.2014.7001517

Navin, G., & Basavaraj, T. N. (2013). Factors affecting performance of CNT FET as a switch in memory cell. In *Proceedings of International Conference on Advanced Nano materials and Emerging Engineering Technologies* (pp. 655-657). IEEE. 10.1109/ICANMEET.2013.6609376

Prabhu, S., & Sarwade, N. (2013). Application of CNTFET as logic gates and its implementation using HSPICE. *International Journal of Modern Engineering Research*, *3*, 3646–3648.

Qi, H., Kim, Y. B., & Choi, M. (2012). A high speed low power modulo 2n+1 multiplier design using carbon-nanotube technology. In *Proceedings of 55th International Midwest Symposium on Circuits and Systems* (pp. 406-409). IEEE.

Raychowdhury, A., Mukhopadhyay, S., & Roy, K. (2004). A Circuit-Compatible Model of Ballistic Carbon Nanotube Field-Effect Transistors. *IEEE Transactions on Computer-Aided Design of Integrated Circuits and Systems*, *23*(10), 1411–1420. doi:10.1109/TCAD.2004.835135

Sahoo, R., Sahoo, S. K., & Sankisa, K. C. (2015). Design of an efficient CNTFET using optimum number of CNT in channel region for logic gate implementation. In *Proceedings of International Conference on VLSI Systems, Architecture, Technology and Applications* (pp. 1-4). IEEE.

Sethi, D., Kaur, M., & Singh, G. (2017). Design and performance analysis of a CNFET-based TCAM cell with dual-chirality selection. *Journal of Computational Electronics*, *16*(1), 106–114. doi:10.100710825-017-0952-4

Shreya, S., & Chandel, R. (2014). Performance analysis of CNTFET based digital logic circuits. In *Proceedings of 2014 Students Conference on Engineering and Systems (SCES)*. Allahabad, India. IEEE. 10.1109/SCES.2014.6880063

Singh, A., Khosla, M., & Raj, B. (2016). Comparative analysis of carbon nanotube field effect transistor and nanowire transistor for low power circuit design. *Journal of Nanoelectronics and Optoelectronics*, *11*(3), 388–393. doi:10.1166/jno.2016.1913

Sinha, S. K., & Chaudhury, S. (2014). Comparative study of leakage power in CNTFET over MOSFET device. *Journal of Semiconductors*, *35*(11), 114–02. doi:10.1088/1674-4926/35/11/114002

Sinha, S. K., & Chaudhury, S. (2015). Analysis of different parameters of channel material and temperature on threshold voltage of CNTFET. *Materials Science in Semiconductor Processing*, *31*, 431–438. doi:10.1016/j.mssp.2014.12.013

Yeole, P., & Padole, D. V. (2015). Design of basic logic gates using carbon nano tube field effect transistor and calculation of figure of merit. In *Proceedings of 7th International Conference on Emerging Trends in Engineering & Technology* (pp. 28-33). IEEE. 10.1109/ICETET.2015.15

Chapter 5
CNTFET-Based Ternary Logic Gates:
Design and Analysis Approach

Suman Rani
I.K.G. Punjab Technical University Jalandhar, Punjab, India & Baba Farid College of Engineering and Technology, Bathinda, India.

Balwinder Singh
ACS Division, Centre for Development of Advanced Computing, Mohali, India

ABSTRACT

In the recent digital designs, there are certain circumstances where energy efficiency and ease is required, and in such situations, ternary logic (or three-valued logic) is favored. Ternary logic is an auspicious supernumerary to the conventional binary (0, 1) logic design techniques as this one is possible to attain straightforwardness and energy efficiency. This chapter deals with the comparative analysis of CMOS and CNTFET-based ternary inverter and universal gates design. The simulation result is analyzed and validated with a Hailey simulation program with integrated circuit (HSPICE) simulations. The average delay and power consumption in CNTFET-based ternary inverter have been reduced by approximately 90.3% and 48.8% respectively, as compared to CMOS-based ternary inverter design. Likewise, delay is reduced by 50% and power gets 99% reduction in ternary CNTFET NAND gate as compared to CMOS-based ternary NAND gat. It is concluded that CNFETs are faster and consume less power compared to CMOS technology.

DOI: 10.4018/978-1-7998-1393-4.ch005

1. INTRODUCTION

Simplicity and energy efficiency are one of the most significant features of up-to-date electronics systems intended for high-performance handy applications. Usually, Logics and algorithms in digital computations arc bascd on thc two-valucd codc, that is, high or low, or yes or no or (0 or 1). On the other hand, in few conditions, we exercise a state in which it is problematic otherwise needless to choose right or false. For illustration, in a transient state (changes from 0 to 1 or 1 to 0), it is difficult to resolve whether the value of the state is 0 or 1. Then ternary logic (three-valued logic) can be used, as an alternative of binary logic (two-valued logic), in which the third value i.e truth is familiarized to signify an indefinite state separately since true and false. Henceforth multivalued Logic substitutes the conventional Boolean characterization of the variable by many values for example Ternary logic (Mukaidono, 1986, p.179).

One of the utmost common MVL areas is Łukasiewicz (1970) and Kleene (1954). Out of these two most predominant MVL fields, Kleene field has been extra widespread and suitable owing to its representative nature and suitable definition for basic operators in the computer discipline world (Azhari et al., 2015). Surrounded by all MVL systems, with e base (e=2.718) principals to the furthermost well-organized implementation of the switching systems. Meanwhile 3 is the nearby amount to e, ternary logic is the superlative substitute for designing the MVL circuits (Moaiyeri et al., 2011, p. 285). Ternary logic offers numerous important benefits over the binary logic in digital designs. For illustration, extra information can be transferred over a specified set of lines, interconnections complexity can be reduced and reduction in components count can be achieved. Moreover, serial and serial-parallel arithmetic actions can be passed faster if three-valued logic is engaged (Balla & Antoniou, 1984, p. 739). Broad investigation on the design and implementation of ternary logic with depletion enhancement complementary metal-oxide-semiconductor (DE-CMOS) integrated circuits can be originatein the technical literature (Heung & Mouftah, 1985, p. 609). Based on MOS technology MVL circuits further categories as Current-mode MVL circuits and another one is Voltage-mode MVL circuits. Owing to the ballistic conduction (high performance) and low I_{OFF} (low power consumption) belongings of Carbon Nanotube Field-effect transistors (CNTFETs) are an appropriate choice toward the buck silicon transistors for the design of logic circuits. The CMOS based multi-threshold design depends on the body effects with changed bias voltages to the base terminal of the transistors (Kumar et al., 2019, p. 3). But in event of the CNTFET based multi-threshold design, the diameter (D) of the CNT determines the threshold voltage (V_{TH}) of the transistor. Thus, a multi-threshold design can be consummate through CNTs with altered diameters in the CNTFETs. Novel techniques based on CNTFET used for ternary logic gate design has been offered and associated with existing resistive load CNTFET logic gate design (Liang et al., 2014, p.695). Amongst these putting into practice approaches based on CNTEFT multiple-valued logic can be named with diode-load and resistive-load (Raychowdhury & Roy, 2005, p. 168). These approaches used single power supply for changing voltage levels (Lin et al., 2009, p. 217). But, in one more method, CNFET-based ternary logic design, has been offered with the dissimilar back biasing voltages and diameters of CNFETs to attain the less power consumption (Nan & Choi, 2010, p. 115). The authors have used together resistive load and transistor load to propose a new circuit based on CNTFET for ternary logic gate implementation. Half adder & multipliers with the transistor load discussed in (Samadi et al., 2011, p. 41). Implemented altogether the likely types of ternary logic, comprising negative, positive and standard ternary logic in a single structure in each of the CNFET based circuits designed (Moaiyeri et al., 2011, p. 285). An Energy and area-efficient Estimated Ternary full adder for nanotechnology withdual-pass transistors,switching logic and transmission gates have been presented in (Bastani et al., 2018, p. 1863). Imprecise computing is the trade-off amongthe accuracy and the energy consumption. The authors (Sharifi et al., 2018, p. 108) investigated the high-performance CNTFET based ternary Full adders simulated at different situationsfor example different supply voltages, different temperature, and operational frequencies. Ternary half adder & Multiplier has been presented using transmission gates and CNT transistors & compared with previous designs for low power VLSI designs (Tabrizchi et al., 2016, p. 423). Energy efficient ternary serial adder has been projected based on CNTFETs, significantlydiminishes the number of devices essential for a ternary serial adder (Moaiyeri et al., 2016, p. 271)

A novel ternary multiplier based on a ternary CMOS (T-CMOS) compact model designed (Kang et al., 2017, p. 25). The authors established low-delay and low-power multi-ternary digit CNTFET-based adder circuits established on unary operators of the multi-valued logic (Srinivasu & Sridharan, 2016, p. 352). The ternary half adder circuit uses binary logic gates is implemented using three groups of 3 AND and 1 OR logic gates. Then ternary half adder circuit practices one input, three-output ternary to the binary decoder to advance the power delay product (Sahoo et al., 2017, p. 368). Ternary Full adder based on CNTFET for low power consumption offered in (Jafarzadehpour & Keshavarzian, 2016, p. 365).

2. REVIEW OF TERNARY LOGIC

If the third value is familiarized to the binary logic, the resulting logic is called Ternary logic functions. Here, 0, 1, and 2 mean the ternary values to signify false (low), undefined (middle), and true (high), correspondingly. To design ternary gates from one Power supply the supposed logic symbols used for the logic level "2" is VDD voltage, for the logic level "0" is the ground power supply and for the logic level "1" is half of the input voltage (1/2 VDD). Any n variable {X1, X2 . . ., Xn } ternary function f(X) is exact as a logic function mapping {0, 1, 2}n to {0, 1, 2}, wherever X = {X1, X2. . ., Xn}. Simple operations of ternary logic can be definite as follows (Lin et al., 2009, p. 217),

Wherever (−): arithmetic subtraction

(+): Logical OR
(•): Logical AND in ternary logic, respectively.

Basic digital designs are the Inverter, NOR gate, and NAND gate. The ternary gates are deliberate according to the convention by equation number 1.

Xi, Xj = {0, 1, 2}

OR Gate: Xi + Xj = max {Xi, Xj}

AND Gate: Xi • Xj = min {Xi, Xj}

INVERTER: $\bar{X}$.= 2− Xi .... (1)

TERNARY INVERTER

The utmost noteworthy elementary gate in ternary gates is the standard (and/or the simple) ternary inverter gate, called STI (Khandelwal & Sharan, 2017, p. 429). Further two ternary inverter gates are NTI (negative) and PTI (positive). The Truth table for STI, NTI and PTI are offered in Table 1 (Samadi et, al, 2017, p.41). So, to design of the ternary inverter total, three inverters are essential and they are a Negative ternary inverter (NTI), a Standard ternary inverter (STI), and a Positive ternary inverter (PTI), if Z_0, Z_1, and Z_2 in (2) are the outputs. The Ternary inverter is a gate with one input X and three outputs represented by Z_0, Z_1, and Z_2 such that given by equation number 2 (Lin et. al, 2009, p. 217).

$$Z_0 = NTI = \begin{cases} 2, & if input X = 0 \\ 0, & if input X \neq 0 \end{cases}.$$

$$Z_1 = STI = \bar{X} = 2 - INPUT \,.$$

$$Z_2 = PTI = \begin{cases} 2, & ifinputX \neq 2 \\ 0, & ifinputX = 2 \end{cases} \ldots\ldots\ldots\ldots\ldots\ldots \quad (2)$$

Ternary NAND and NOR Gates

The ternary NAND and NOR gates are two multiple input operators that are used in ternary logic. The operation of two-entry ternary NAND and NOR gates are definite by equations number 3 & 4, correspondingly (Lin et al., 2012, p. 1019).

$$Y_{NAND} = \overline{\min\{\, X1,\ X2\}} \ldots\ldots\ldots\ldots \quad (3)$$

$$Y_{NOR} = \overline{\max\{\, X1,\ X2\}} \ldots\ldots\ldots\ldots \quad (4)$$

The truth table used for the ternary NAND & NOR gate having two inputs X1 and X2 is shown in Table II.

3. CARBON NANOTUBE FET

Carbon Nanotubes (CNTs) was exposed in 1991 by Sumio lijima, Japanese physicist, by coincidence when performing experiments on Carbonium (Chowdhury et al., 2012, p. 28). CNTs, the allotropes of carbon which belongs to the fullerenes family, are the Nano-scaled tube formed by the rolled sheets of graphite. The length of CNT varieties from fewer than a micron to several millimeters or even in centimeters (Moaiyeri et al., 2013, p.167). Most fascinating property of CNT is their unique –quasi-one-dimensional Nanostructures. The most eye-catching feature of these

Table 1. Truth table of ternary inverter STI, PTI and NTI

Input X	NTI (Z_0)	STI (Z_1)	PTI (Z_2)
0	2	2	2
1	0	1	2
2	0	0	0

Table 2. Truth table for NAND and NOR Gates

Input (X1)	Input (X2)		
0	0	2	2
1	0	2	1
2	0	2	0
0	1	2	1
1	1	1	1
2	1	1	0
0	2	2	0
1	2	1	0
2	2	0	0

structures is high electrical and thermal conductivity, high tensile strength, high ductility and chemical stability which leads to the future applications in Nano-Electronics. Nanotubes are categorized according to the number of cylinders, into single-walled nanotubes (SWNT) and multi-walled nanotubes (MWNT) (Kumara et al., 2018). SWCNT structures contain of a solitary layer of carbon sheet, further classified into three types such as armchair, chiral, and zigzag, depending on the chiral vector of the CNTs, which is firm by (n, m) indices, if n = m, the SWCNTs are named as arm-chair, if m = 0, these are so-called zig-zag, and the respite are known as chiral (Bacon, 1960, p. 283). Another kind of classification of CNTs depends upon the chiral number (n, m) which defines the tube as conducting or metallic. If (n-m) is divisible by 3, the tube is metallic otherwise semiconducting (Singh et al., 2016, p. 5381). A carbon nanotube Field-effect transistor is referred to as FET that use single or multiple CNT as a channel among the source and drain terminal. Semiconducting SWCNTs are used as the channel in the transistors grounded on the carbon nanotubes (Bastani et al., 2018, 1863). Figure 1 signifies the descriptive structure of a CNTFET.

Chirality defines the basic property of CNT like diameter, bandwidth and threshold voltage. The diameter of CNT determines the bandgap energy of the tube shown in equation number 5 (Lin et al., 2012, p. 1019).

$$D_{CNT} = \sqrt{3}\frac{a0}{\pi}\sqrt{m^2 + n^2 + mn}$$

or

Figure 1. Descriptive structure of CNTFET

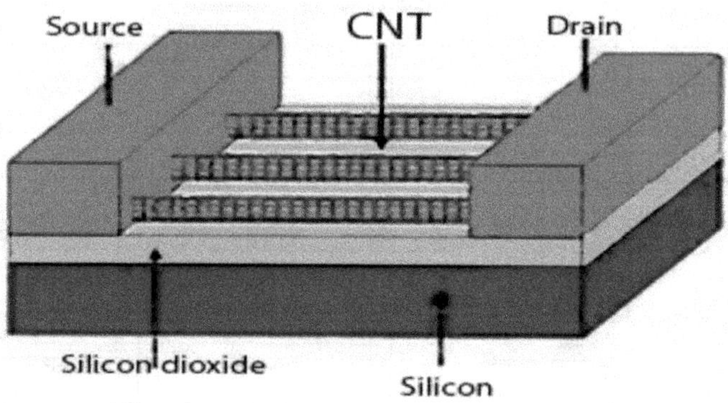

$$D_{CNT} = 0.0783\sqrt{m^2 + n^2 + mn} \quad \ldots\ldots\ldots\ldots\ldots \quad (5)$$

Here a0 = 0.142nm

Threshold voltage (Vth)is well-defined as the least voltage mandatory to turn on the device. Threshold voltage (Vth) for intrinsic CNTFET is approached to the 1st ordered as the half band gap given by equation number 2 and is the inverse function of diameter given by equation number 6 (Lin et al., 2012, p. 1019).

$$V_{th} = \frac{E_g}{2c} = \frac{\sqrt{3}}{3} a \frac{V\pi}{(eD_{CNT})} = \frac{0.43}{D_{CNT}(\mathrm{nm})} \quad \ldots\ldots\ldots\ldots\ldots\ldots\ldots\ldots \quad (6)$$

Where a= 2.49 °A is Carbon to carbon space, Vπ = 3.033ev is carbon π-π bond energy in close-fitting bonding model, e is unit electron charge and D_{CNT} is the diameter of the nanotube. As the chirality vector alterations, the threshold voltage of the CNTFET will likewise change. Assume that m in the chirality vector is permanently zero; then, the ratio of the threshold voltages of two CNTFETs by different chirality vectors is agreed as equation number 7 (Lin et. al 2012, p. 1019).

$$\frac{V_{th1}}{V_{th2}} = \frac{D_{CNT2}}{D_{CNT1}} = n_2 / n_1 \quad \ldots\ldots\ldots\ldots\ldots\ldots\ldots\ldots \quad (7)$$

The MWCNT structures can be understood by the concept of packagingnumerous graphite layers into a concentric seamless cylinder. There are two models can be used to designate the structure of MWCNTs, namely Russian Doll Model and Parchment model (Bacon, 1960, 283).

Basic standard operation of CNFET is the similar as MOSFET where electrons are conveyed by source terminal and drain terminal will accumulate these electrons. Gatc tcrminal pancls currcnt intcnsity in thc transistor channcl and if no voltage is applied at the gate terminal then the transistor is in off state.

4. CNTFET Device Structure

CNTFETs are the new kind of the device which replaces the conventional MOSFET channel with the carbon nanotube material (Avouris et. al, 2005, p. 227). There are basic three types of structures namely Schottky Barrier (SB-CNFET), Conventional (C-CNFET) and Tunnel (T- CNFET) discussed below (Singh et. al 2016, p. 074001).

The current- voltage characteristics of CNTFET depend on its diameter (d), as the bandgap energy is inversely proportional to diameter (d). Each device concept has been discussed briefly as below:

Schottky Barrier Cntfet: In this device source and drain contacts are directly connected to gate controlled nanotube. The entire channel is intrinsic with the Length (LG) as shown in Figure 2. SB-CNFET is a tunneling transistor and performs on the principal of direct tunneling via Schottky- Barrier at source/ drain channel junction (Singh et al., 2015, p. 552).

As the entire channel is intrinsic, On-current (Ion) for both positive and negative gate voltages is observed which are termed as ambipolar characteristics. These ambipolar characteristics are obtained due to hole injection from negative gate voltage and electron injection from positive gate voltage, with distinct drain

Figure 2. Schottky barrier (SB- CNFET)

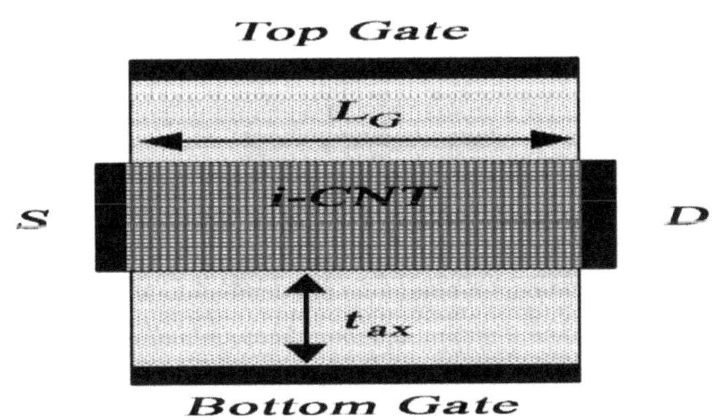

voltage dependence (VDS). SB-CNFET has strong ambipolar attributes that limit the usage of this device in convention CMOS architecture. An important aspect of SB- CNFET is that transport inside the nanotube is ballistic and the drain voltage drops correspondingly exclusively at the electrodes. The gate voltage alters the tunneling probability, thus the current through the device. The Inverse sub-threshold slope $S=dV_{GS}/dLog_{10}I_d$ is always greater than 60mV/dec. Due to these points, it is necessary to search for alternative devices and their concept (Appenzeller et al., 2005, p. 2568).

Conventional CNFET: The structure of C- CNFET is similar to the conventional N-type and P-Type MOSFET, with the difference of undoped gated region as desired for ultra- thin body devices and the channel is formed by CNT. The p/i/p or n/i/n doping profile are introduced along the channel by gating only the undoped portion of the tube and highly doped areas are close to the contact (Figure 3). The main advantage of C-CNFET is that unipolar device characteristics with low off- current can be obtained whereas, the major disadvantage of C-CNFET is the electron-pile up occur in the gated region of p-Type C-CNFET and holes pile up occur in the gated region of n-Type C-CNFET making it a leaky device. Due to this problem of charge pile up and in order to make gate controlled tunneling a relevant approach for device applications C-CNFET is not the appropriate type (Appenzeller et al., 2005, p. 2568).

Tunnel-CNFET: In this type of CNTFET, the doping profile is p/i/n or n/i/p instead of p/i/p or n/i/n as in the case of C-CNFET as shown in Figure 4 (Appenzeller et al., 2005, p. 2568).Due to the tunneling behavior of the device, the problem of charge piles up in C-CNFET is also eliminated and the tunneling is controlled by gate voltage so-called gate controlled tunneling CNFET (Rani & Singh, 2018, p. 76). Here the Inverse subthreshold slope, $S=dV_{GS}/dLog_{10}I_d$, is always smaller than 60mV/dec. T- CNFET is very appropriate for Ultra low power and subthreshold circuit designing.

Figure 3. C-CNFET (n- Type)

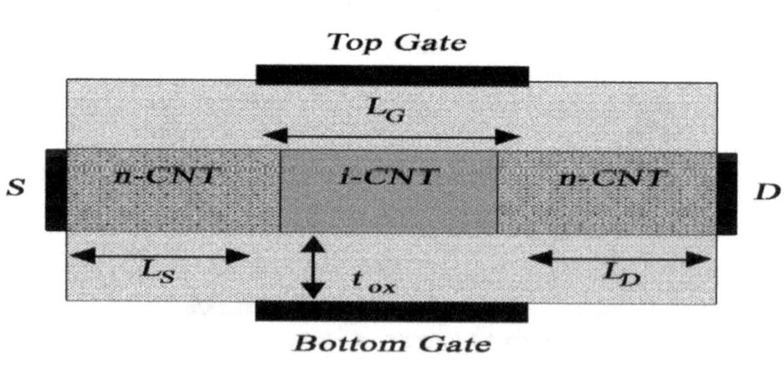

Figure 4. T –CNFET (p-i-n type)

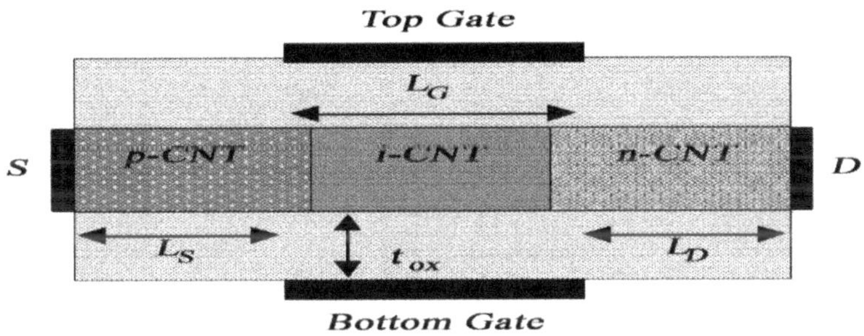

5. DESIGN OF CMOS AND CNTFET BASED TERNARY LOGIC GATES

CMOS and CNTFET Based Ternary Inverter Design

In the ternary inverter, we need to obtain three output values viz. 2, 1 and 0 corresponding's to input values 0, 1 and 2 respectively. So, two transistors and one voltage source are employed for the design of CMOS ternary inverter design, as shown in the schematic of figure 5 (Shin et al., 2015, p. 2396).

Just by varying the sizing of transistors from this same architecture all three styles of ternary CMOS inverter STI, PTI, NTI style can be achieved. The resistance of the channels can be changed by varying the aspect ratio (W/L) of the PMOS and NMOS transistors (Gaikwad & Deshmukh, 2015. P. 1).

Similarly, a CNTFET based standard ternary inverter design is simulated. The circuit has six CNTFETs in which two NCNTFETs and four PCNTFETs are employed shown in figure 6 (Avouris et al., 2005, p. 227). In the circuits of ternary logic designs, the variable threshold voltage is required. Therefore, CNTs with dissimilar chirality vectors are used to achieve multiple threshold voltages.

All three styles STI, PTI and NTI of the CNTFET based ternary inverter is shown in Figure 6. The chirality vectors of each transistor are indicated in Figure 6. The equivalent diameter for (19, 0) and (10, 0) is 1.487nm and 0.783nm while the threshold voltage is 0.293 and 0.557 correspondingly (Avouris et al., 2005, p. 227).

For implementing NTI, the Transistor XP2 (P-CNFET) with High-Vth and Transistor XN2 (N-CNFET) with low- Vth transistors are used. Similarly, for implementing PTI Transistor XP1(P-CNFET) with the low- Vth and high- Vth, transistors XN1(N-CNFET) are employed. The outputs of NTI and PTI gates are linked as one through an N-CNFET (XN3 Transistor) and a P-CNFET (XP3 Transistor) having Chirality of (19, 0) (Avouris et al., 2005, p. 227). Therefore, the voltage of the joining node will be as mentioned by equation number 8

Figure 5. CMOS based ternary inverter design (Shin et al., 2015, p. 2396)

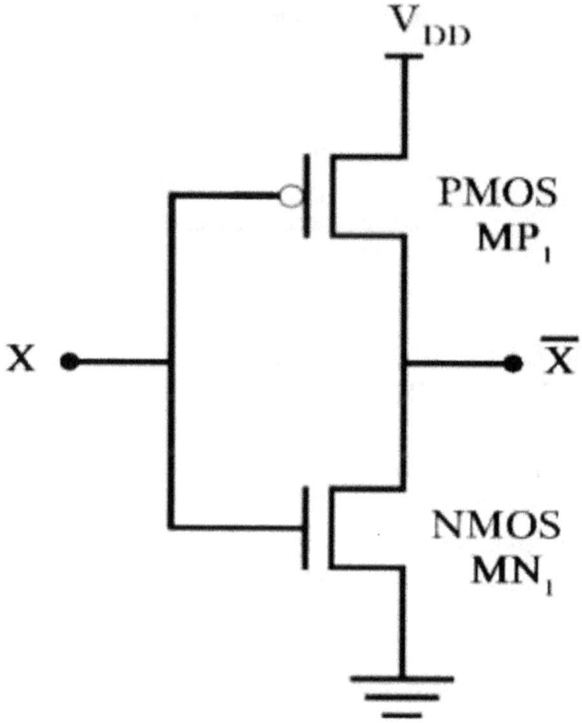

Figure 6. CNTFET based Ternary Inverter Design (Avouris et al., 2005, p. 227)

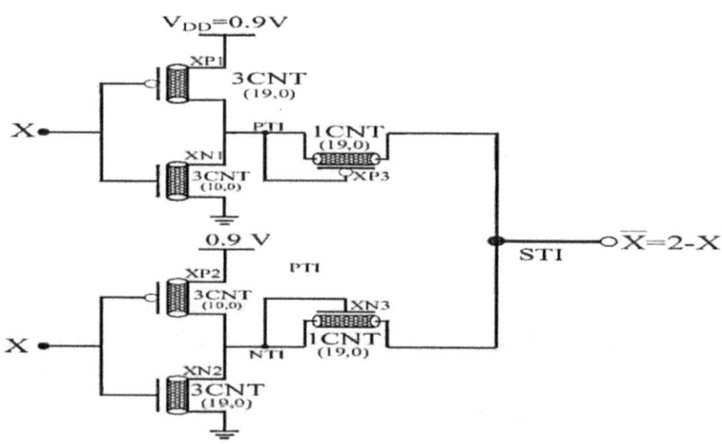

$$V_{STI} = \frac{V_{NTI} + V_{PTI}}{2} \text{........................} \quad (8)$$

CMOS and CNTFET Based Ternary NAND and NOR Gate Design:

Figure 7 shows the diagrams of CMOS based ternary NAND and NOR gate. It is recognized by connection of a CMOS based transmission gate on the common drain output of a binary CMOS NAND and NOR gate (Mano & Ciletti, 2012).

The design of CNTFET based logic gates is different from that of CMOS based design. All the three styles STI, PTI and NTI of CNTFET based ternary NAND & NOR gate is exposed in figure 8 (a) and 8 (b) using 10 transistors.

6. SIMULATION RESULTS AND DISCUSSION

In this work, High-performance CNTFET based Ternary universal gates designs compared with bulk CMOS based designs at 32 nm technology (Stanford University CNFET Model, Predictive Technology Model). Synopsis HSPICE simulator is used for simulation and comparison. All the circuits are simulated by the power supply of 0.9 V. Transient analysis has been done for the period of 1 to 100ns. Table 3 and Table 4 represents the parameters for CNFET model and bulk CMOS based simulations respectively:

Here CMOS implementation used 0.9 V supply and sizing is done at every stage for obtaining wanted aspect ratio. Figure 9 (a) and (b) displays the transient response of CMOS and CNTFET based Ternary inverter designs.

The performance parameters of the two technology-based basic gate designs have been tabulated in table no 5. It is perceived that the delay, power and PDP have reduced in CNTFET based inverter as equated to CMOS based ternary inverter.

Figure 7. CMOS based ternary gate (a) NAND gate (b) NOR gate (35)

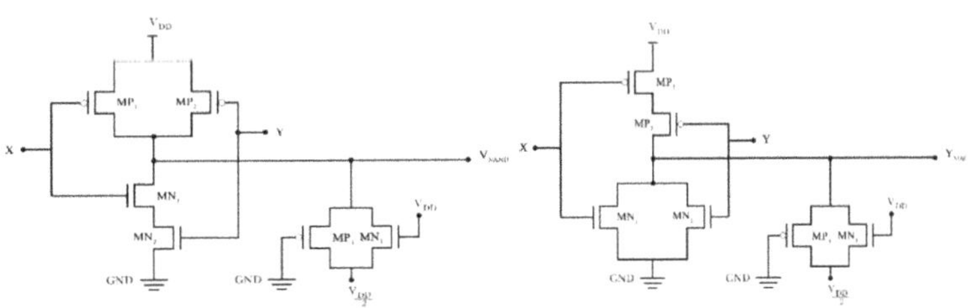

Figure 8. CNTFET based ternary gate (a) and (b) (Avouris et al., 2005, p. 227)

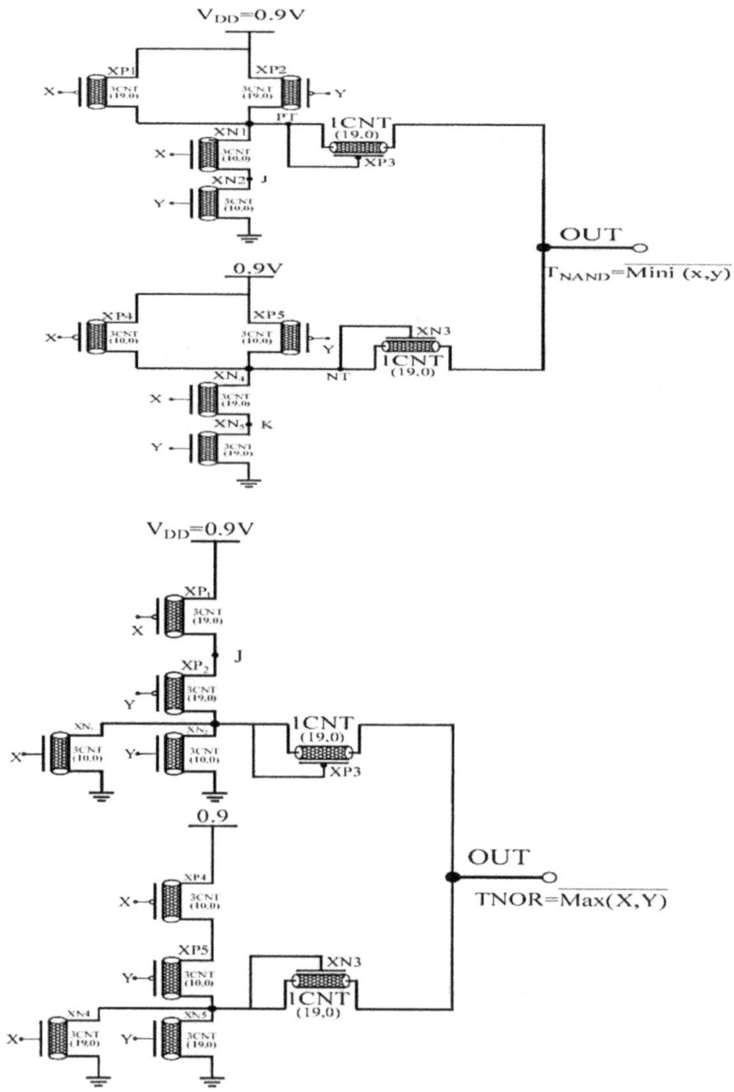

Hence it can be said that the performance of CNTFET based Inverter, NAND Gate and NOR Gate circuits have great enrichment in terms of delay, power consumption and PDP as related to CMOS based implementation.

Figure 10 (a) and (b) shows the transient response of CMOS and CNTFET based Ternary NAND gate designs.

Figure 11 (a) and (b) shows the transient response of CMOS and CNTFET based Ternary NOR gate designs.

Table 3. Key parameters of CNFET Model for simulation

Parameter	Ternary CNFET
Physical Channel length	32nm
Power supply (VDD)	0.9V
CNT count	01, 03
Chiral Numbers (m, n)	(m=19, n=0), (ma=10, na=0)
CNT Diameter	1.487nm, 0.783nm
Threshold Voltage	0.293V, 0.398V
Coupling capacitance between channel & substrate (Cb)	40e-12

Table 4. Key parameters of Bulk CMOS Model for simulation

Parameter	Ternary CMOS
Physical Channel length	32nm
Power supply (VDD)	0.9V
The gate to Oxide thickness (TOX)	1.6nm
The source and drain resistance per unit channel length (RDS)	190 Ω-µm

Figure 9. Transient analysis of ternary inverter (a) CMOS based (b) CNTFET based

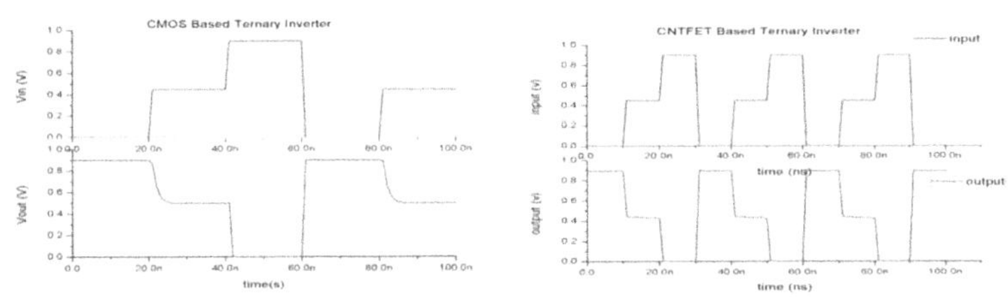

Table 5. Performance comparison of CNTFET and CMOS based basic gate designs

Design →	Ternary Inverter		Ternary NAND Gate		Ternary NOR Gate	
Parameter↓	CMOS Based	CNTFET Based	CMOS Based	CNTFET Based	CMOS Based	CNTFET Based
Delay(s)	2.04E-07	1.98E-08	1.96E-08	9.75E-09	1.04E-07	1.98E-08
Average Power(W)	5.06E-08	2.59E-08	1.96E-05	1.91E-08	5.42E-06	1.69E-06
PDP(J)	1.03E-14	5.11E-16	3.85E-13	1.86E-16	5.64E-13	3.34E-14

Figure 10. Transient analysis of ternary NAND gate (a) CMOS based (b) CNTFET based

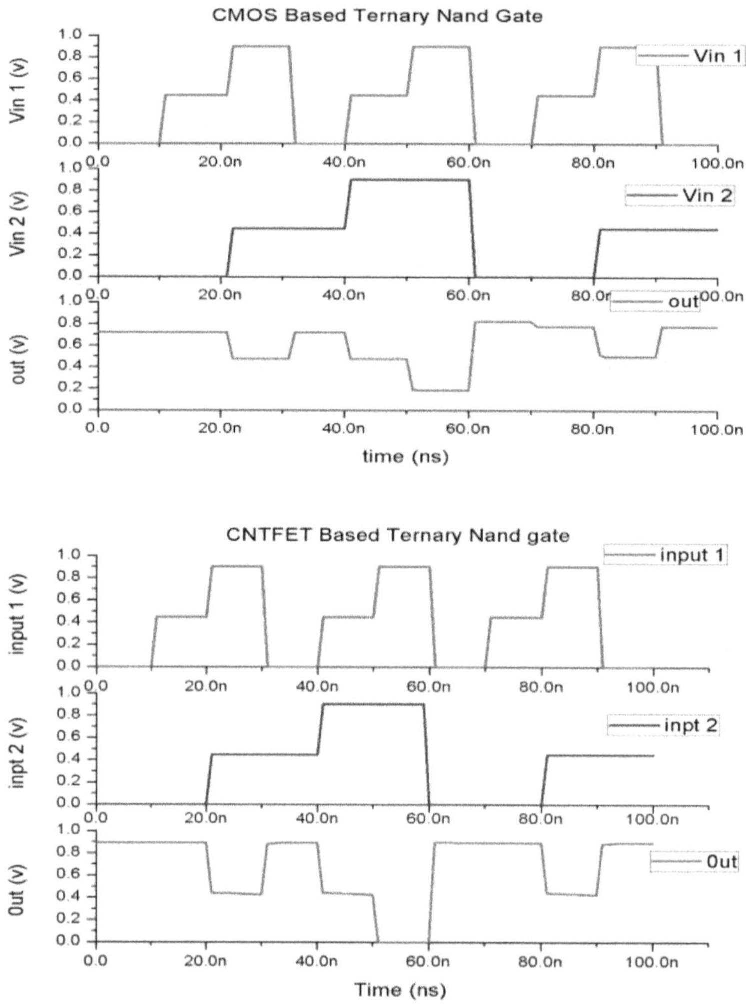

7. CONCLUSION

The conclusion of the Ternary research work review may have several applications in the fields of VLSI Designs, ultra-thin body devices and will have an impact on reduced pin count and increased serial data transfer rates make it better when compared to binary designs. Multiple Design methods for low power VLSI design and the various techniques that will be used to reduce power are taken into account as a further extension of this work. Reduction in power, Pin count and increased

Figure 11. Transient analysis of ternary NOR gate (a) CMOS based (b) CNTFET based

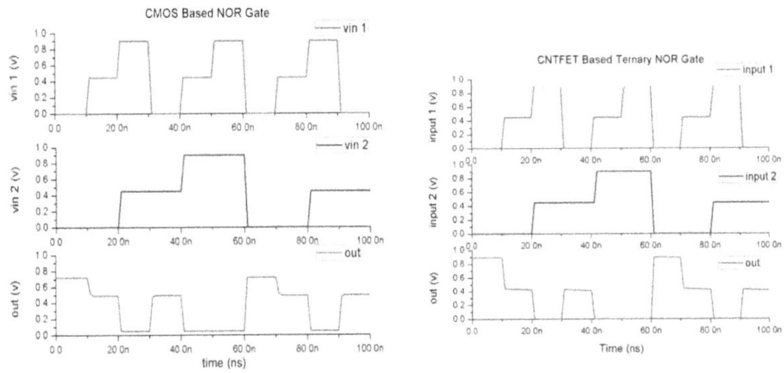

serial data transfer rates will help in the implementation of this research for high-performance applications. The ternary inverters and Universal Gates based on CMOS and CNT have been simulated at 32nm using HSPICE. It is perceived that amongst the devices compared T-CNFET may be considered as vital Nano -scale device for high-performance applications. Furthermore, it is observed that as compared to CMOS, CNT is a better alternative to design ternary logic based circuits because of its superior characteristics.

8. SUMMARY

In this chapter a brief introduction about Carbon Nanotube & device structures based on that has been presented. Chapter summarize that ternary logic is better because it can process more information as compared to binary. Correspondingly, CNT is a good option for designing ternary logic based circuits. Average delay and power consumption in CNTFET based Ternary Inverter have been reduced by approximately 90.3% and 48.8% correspondingly, as compared to CMOS based Ternary Inverter design. Similarly, Delay is reduced by 50% and power gets 99% reduction in Ternary CNTFET NAND Gate as compared to CMOS Based Ternary NAND Gate. Furthermore, in the case of a ternary CNTFET NOR Gate, decrease in the delay is around 81% and that of power is reduced by 69% as associated to CMOS Based Ternary NOR Gate.

REFERENCES

Appenzeller, J., Lin, Y. M., Knoch, J., Chen, Z., & Avouris, P. (2005). Comparing carbon nanotube transistors-the ideal choice: A novel tunneling device design. *IEEE Transactions on Electron Devices*, *52*(12), 2568–2576. doi:10.1109/TED.2005.859654

Avouris, P., Radosavljević, M., & Wind, S. J. (2005). Carbon nanotube electronics and optoelectronics. In *Applied Physics of Carbon Nanotubes* (pp. 227–251). Berlin: Springer. doi:10.1007/3-540-28075-8_9

Azhari, R., Shahhoseini, A., & Aghaei-liavali, F. (2015). Stability Improvement and consumption power of CNTFET-based ternary memory ell. *2015 IEEE International Conference on Research in Computational Intelligence and Communication Networks (ICICRCICN, 2015).*

Bacon, R. (1960). Growth, structure, and properties of graphite whiskers. *Journal of Applied Physics*, *31*(2), 283–290. doi:10.1063/1.1735559

Balla, P. C., & Antoniou, A. (1984). Low power dissipation MOS ternary logic family. *IEEE Journal of Solid-State Circuits*, *19*(5), 739–749. doi:10.1109/JSSC.1984.1052216

Bastani, N. H., Moaiyeri, M. H., & Navi, K. (2018). An energy-and area-efficient approximate ternary adder based on CNTFET switching logic. *Circuits, Systems, and Signal Processing*, *37*(5), 1863–1883. doi:10.100700034-017-0627-1

Chowdhury, A. H., Akhter, N., & Al Faisal, A. (2012). Performance Analysis and Development of Self-Consistent Model of Carbon Nanotube Field Effect Transistor (CNTFET). In *Proc. of the Global Engg, Sci. and Tech. conf* (pp. 28-29). Academic Press.

Gaikwad, V. T., & Deshmukh, P. R. (2015, January). Design of CMOS ternary logic family based on single supply voltage. In *2015 International Conference on Pervasive Computing (ICPC)* (pp. 1-6). IEEE. 10.1109/PERVASIVE.2015.7087114

Heung, A., & Mouftah, H. T. (1985). Depletion/enhancement CMOS for a lower power family of three-valued logic circuits. *IEEE Journal of Solid-State Circuits*, *20*(2), 609–616. doi:10.1109/JSSC.1985.1052354

Jafarzadehpour, F., & Keshavarzian, P. (2016). Low-power consumption ternary full adder based on CNTFET. *IET Circuits, Devices & Systems*, *10*(5), 365–374. doi:10.1049/iet-cds.2015.0264

Kang, Y., Kim, J., Kim, S., Shin, S., Jang, E. S., Jeong, J. W., . . . Kang, S. (2017, May). A novel ternary multiplier based on ternary CMOS compact model. In *2017 IEEE 47th International Symposium on Multiple-Valued Logic (ISMVL)* (pp. 25-30). IEEE. 10.1109/ISMVL.2017.52

Khandelwal, M., & Sharan, N. (2017). Design and Performance Comparison of CNTFET-Based Binary and Ternary Logic Inverter and Decoder With 32 nm CMOS Technology. In *Advances in Computer and Computational Sciences* (pp. 429–437). Singapore: Springer. doi:10.1007/978-981-10-3770-2_40

Kleene, S. C. (1954). *Introduction to metamathematics*. Academic Press.

Kumar, H., Srivastava, S., & Singh, B. (2019). Comparative Analysis of 6T, 7T Conventional CMOS and CNTFET Based SRAM Cell Design. *Advanced Science Engineering in Medicine*, *11*(1-2), 3–10.

Kumari, A., Rani, S., & Singh, B. (2018, July). I-V Characteristics analysis of Carbon Nanotube Field effect transistor and Graphene Nanotube Field-effect transistor. *IEEE Conference KIIT Bhubaneswar*.

Lee, S. Y., Kim, S., & Kang, S. (2019, May). Ternary Logic Synthesis with Modified Quine-McCluskey Algorithm. In *2019 IEEE 49th International Symposium on Multiple-Valued Logic (ISMVL)* (pp. 158-163). IEEE. 10.1109/ISMVL.2019.00035

Liang, J., Chen, L., Han, J., & Lombardi, F. (2014). Design and evaluation of multiple valued logic gates using pseudo N-type carbon nanotube FETs. *IEEE Transactions on Nanotechnology*, *13*(4), 695–708. doi:10.1109/TNANO.2014.2316000

Lin, S., Kim, Y. B., & Lombardi, F. (2009). CNTFET-based design of ternary logic gates and arithmetic circuits. *IEEE Transactions on Nanotechnology*, *10*(2), 217–225. doi:10.1109/TNANO.2009.2036845

Lin, S., Kim, Y. B., & Lombardi, F. (2012). Design of a ternary memory cell using CNTFETs. *IEEE Transactions on Nanotechnology*, *11*(5), 1019–1025. doi:10.1109/TNANO.2012.2211614

Łukasiewicz, J. (1970). *Selected Works, L.* Borkowski.

Moaiyeri, M. H., Doostaregan, A., & Navi, K. (2011). Design of energy-efficient and robust ternary circuits for nanotechnology. *IET Circuits, Devices & Systems*, *5*(4), 285–296. doi:10.1049/iet-cds.2010.0340

Moaiyeri, M. H., Mirzaee, R. F., Doostaregan, A., Navi, K., & Hashemipour, O. (2013). A universal method for designing low-power carbon nanotube FET-based multiple-valued logic circuits. *IET Computers & Digital Techniques*, *7*(4), 167–181. doi:10.1049/iet-cdt.2013.0023

Moaiyeri, M. H., Nasiri, M., & Khastoo, N. (2016). An efficient ternary serial adder based on carbon nanotube FETs. *Engineering Science and Technology, an International Journal, 19*(1), 271-278.

Morris Mano, M., & Ciletti, M. D. (2012). *Digital Design: With an Introduction to the Verilog HDL*. Academic Press.

Mukaidono, M. (1986). Regular ternary logic functions? ternary logic functions suitable for treating ambiguity. *IEEE Transactions on Computers*, *C-35*(2), 179–183. doi:10.1109/TC.1986.1676738

Nan, H., & Choi, K. (2010, November). Novel ternary logic design based on CNFET. In 2010 International SoC Design Conference (pp. 115-118). IEEE. doi:10.1109/SOCDC.2010.5682960

Predictive Technology Model. (n.d.). Retrieved from http://ptm.asu.edu/modelcard/LP/32nm_LP.pm

Rani, S., Kaur, G., & Lakha, B. S. (2018). Ternary Logic Design Approach-From CMOS to CNTFET. *Journal of Nanoscience Nanoengineering and Applications*, *8*(2), 20–30.

Rani, S., & Singh, B. (2018). Investigation of Schottky barrier, conventional and tunnel carbon nanotube field effect transistor for low power design. *Journal of Nanoelectronics and Optoelectronics*, *13*(1), 76–82. doi:10.1166/jno.2018.2169

Raychowdhury, A., & Roy, K. (2005). Carbon-nanotube-based voltage-mode multiple-valued logic design. *IEEE Transactions on Nanotechnology*, *4*(2), 168–179. doi:10.1109/TNANO.2004.842068

Sahoo, S. K., Akhilesh, G., Sahoo, R., & Muglikar, M. (2017). High-performance ternary adder using CNTFET. *IEEE Transactions on Nanotechnology*, *16*(3), 368–374. doi:10.1109/TNANO.2017.2649548

Samadi, H., Shahhoseini, A., & Aghaei-liavali, F. (2017). A new method on designing and simulating CNTFET_based ternary gates and arithmetic circuits. *Microelectronics Journal*, *63*, 41–48. doi:10.1016/j.mejo.2017.02.018

Sharifi, F., Panahi, A., Moaiyeri, M. H., Sharifi, H., & Navi, K. (2018). High performance CNFET-based ternary full adders. *Journal of the Institution of Electronics and Telecommunication Engineers, 64*(1), 108–115. doi:10.1080/03772063.2017.1338973

Shin, S., Jang, E., Jeong, J. W., Park, B. G., & Kim, K. R. (2015). Compact design of low power standard ternary inverter based on OFF-state current mechanism using nano-CMOS technology. *IEEE Transactions on Electron Devices, 62*(8), 2396–2403. doi:10.1109/TED.2015.2445823

Singh, A., Khosla, M., & Raj, B. (2015, October). Comparative analysis of carbon nanotube field effect transistors. In *2015 IEEE 4th Global Conference on Consumer Electronics (GCCE)* (pp. 552-555). IEEE. 10.1109/GCCE.2015.7398601

Singh, A., Khosla, M., & Raj, B. (2016). Circuit compatible model for electrostatic doped Schottky barrier CNTFET. *Journal of Electronic Materials, 45*(10), 5381–5390. doi:10.100711664-016-4743-7

Singh, A., Saini, D. K., Agarwal, D., Aggarwal, S., Khosla, M., & Raj, B. (2016). Modeling and simulation of carbon nanotube field effect transistor and its circuit application. *Journal of Semiconductors, 37*(7), 074001. doi:10.1088/1674-4926/37/7/074001

Srinivasu, B., & Sridharan, K. (2016). Carbon nanotube FET-based low-delay and low-power multi-digit adder designs. *IET Circuits, Devices & Systems, 11*(4), 352–364. doi:10.1049/iet-cds.2016.0013

Stanford University CNFET Model. (n.d.). Retrieved from http://nanostanford.edu/models.phpwebsite

Stanford University CNTFET Model Website. (2008). *Stanford University*. Available: http://nano.stanford.edu/model. php? id=23

Tabrizchi, S., Azimi, N., & Keivan, N. (2016). Design a novel ternary half adder and multiplier based on carbon nano-tube field effect transistors (CNTFETs). *Frontiers, 1*, 423–433.

KEY TERMS AND DEFINITIONS

Carbon Nanotube: CNTs, the allotropes of carbon that belong to the fullerenes family, are the Nano-scaled tube formed by the rolled sheets of graphite.

Carbon Nanotube Field-Effect Transistor: A carbon nanotube field-effect transistor is referred to as FET. It utilizes one or multiple CNTs as a channel between the source and drain terminal.

Delay: Is the length of time that starts when the input to a logic gate becomes stable and valid to change to the time that the output of that logic gate is stable and valid to change.

Multi-Valued Logic Design: To enhance the performance of CMOS technologies, MVL modules have been inserted in binary logic ICs. Based on MOS technology, MVL circuits further categories as current-mode MVL circuits and another one is voltage-mode MVL circuits.

Ternary CMOS: A ternary CMOS is a transistor that uses ternary logic (three possible values) as an alternative of the more widely held binary system ("Base 2") in its calculations.

Chapter 6
Low Power, High Performance CNTFET-Based SRAM Cell Designs:
Design and Analysis of 6T CNTFET SRAM Cells

B. K. Madhavi
Siddhartha Institute of Engineering and Technology, Hyderabad, India

Rajendra Prasad Somineni
https://orcid.org/0000-0002-3178-1239
Vallurupalli Nageswara Rao Vignana Jyothi Institute of Engineering and Technology, Hyderabad, India

ABSTRACT

The main objective of this chapter is to provide high-performance, low-power solutions for VLSI system designers. As technology scales down to 32nm and below, the present CMOS technology has to face the scaling limit, such as the increased leakage power, SCEs, and so on. To overcome these limits, the researchers have experimented on other technologies, among which a CNT technology-based device called CNTFET has been evaluated as one of the promising replacements to CMOS technology. In any digital systems, memory is an integral part, and it is also the largest constituent. SRAM is a widely used memory. In today's ICs, SRAM is going to occupy 60-70% of the total chip area. In this connection, this chapter describes the design of CNTFET-based 6T SRAM cell using circuit-level leakage reduction techniques, named sleep transistor, forced stack, data-retention sleep transistor, and stacked sleep.

DOI: 10.4018/978-1-7998-1393-4.ch006

1. MOTIVATION

Previously, the prime objectives of the VLSI designer were area, performance, cost and reliability, but power consideration was mostly of only secondary importance. In the present scenario the power consumption became a major concern. Continuous decrease in the feature size and the consequent increase in chip density and operating frequency have made power consumption a major concern in VLSI design (Meindl, 1995).

For a CMOS circuit, the total power dissipation consists of dynamic power and static power. In the standby mode, the power dissipation is due to the leakage current. Dynamic power has two components, called short circuit power and switching power. One is due to charging and discharging of load capacitance called the switching power. The other is due to the non-zero rise and fall time of input waveforms called short circuit power. Leakage power is the consumed power that is due to the current flowing through the transistors of a circuit that are in the cut-off state.

Reduction of supply voltage reduces the power dissipation; this is because the power dissipation has quadratic relationship with the supply voltage. But reducing the supply voltage degrades the performance. In order to satisfy the high performance requirements, V_{TH} has to be scaled. Unfortunately, such scaling leads to a dramatic increase in leakage current, which becomes a new concern for low power and high performance circuit designs (Chandrakasan, Sheng & Brodersen, 1992). Earlier in micrometer technologies the dynamic power is the predominant component of the total power but in nanometer technologies leakage power is the dominant component. Starting at 70nm semiconductor technology, leakage power will contribute with 50% to total circuit's power (Pant, De & Chstterjee, 1997).

As technology scales down to 32nm and below, the bulk CMOS technology has reaching the scaling limit. This scaling limit is due to exponential increase of leakage power, the increased short-channel effects, severe process variations, high power density etc. To overcome this scaling limit, various types of materials have been experiment by the researchers. Among these alternative devices MOSFET-like CNTFET device has been evaluated as one of the hopeful replacements in the nanoscale electronics. The reason that makes CNTFET is a promising device is that, it is compatible with high dielectric constant materials, a unique 1-D band-structure that restrains back-scattering and makes near-ballistic operation. Using the CNTFET, a high-k gate oxide can be deployed for lower leakage currents while keeping the on-current drive capability with MOSFET. CNTFET has lower short-channel effect, hence low Off-current and a higher sub-threshold slope than Si-MOSFET (Akturk, Pennington, Goldsman, & Wickenden, 2007).

For any digital system design a memory is an important integral part. SRAM is the widely used memory. It is going to occupy as much as 60-70% of the total chip area in today's ICs and about 75%-85% of the transistor count in some IC products (de Alba Rosano & Garcia-Garcia, 2006). So it is important to reduce leakage power in SRAM, reduction of a leakage power in SRAM leads to reduction of overall chip leakage power.

The first CNTFET was reported in the year 1998. A great progress has been made during the past years in all the areas of CNTFET science and technology, including circuits, devices and materials (Patil, Lin, Zhang, Wong, & Mitra, 2009). The record-breaking electrical and physical properties of CNTs make it attractive for use in applications such as Memories. The CNTFET has been presented in SRAM cell design to achieve low-power as an alternative material for CMOS technology (Sheng Lin, Yong-Bin Kim & Fabrizio Lombardi, 2008 & Wei Wang, & Ken Choi, 2010). Therefore the design of a low power CNTFET SRAM cells for Digital system is very important.

1.1. Objectives of the Chapter

The main objective of this chapter is to provide low power solutions for VLSI circuit designers. Especially, this chapter focuses on leakage power reduction solutions. In this chapter, a circuit structures called Sleep Transistor, Forced Stack, Data-Retention Sleep Transistor and Stacked Sleep techniques are presented. These techniques are applied to a CNTFET based 6T SRAM cell.

The specific objectives of present chapter are given below in chronological order:

1. A study on Low leakage power CMOS/CNTFET digital circuit techniques.
2. Design and analysis of 6T CNTFET SRAM cell using circuit-level leakage reduction techniques, named as sleep transistor, forced stack, data-retention sleep transistor, and stacked sleep. Comparison of this 6T CNTFET SRAM cell with CMOS 6T SRAM cell in terms of leakage power, dynamic power, read delay, write delay and static noise margin (SNM) to assess its performance.

1.2 Review of CNTFET and SRAM

1.2.1 Carbon NanoTube Field Effect Transistor (CNTFET)

CNTFET is a three-terminal device consisting of a source, drain and gate. CNTFET is nothing but a FET, but it uses CNT as channel. In contrast MOSFETs uses silicon as their channel. CNT acts as conducting channel between source and drain. CNT

Figure 1. CNTFET (a) Three dimensional view (b) Top view
(a) (b)

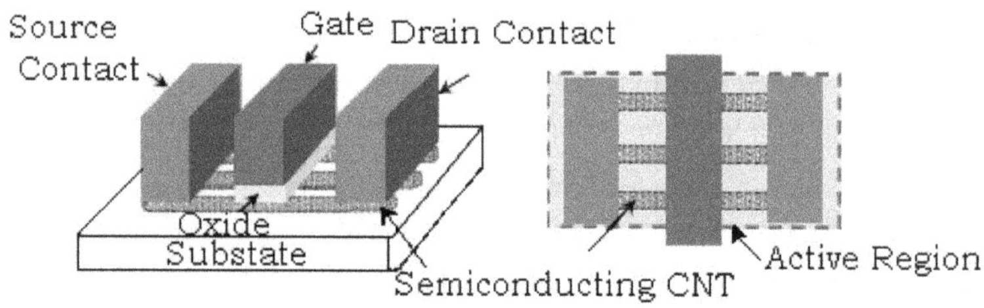

channel is turned ON or OFF electrically via the gate as shown in Figure 1 (Sheng Lin, Yong-Bin Kim & Fabrizio Lombardi, 2008).

The V-I characteristics of CNTFET are similar to those of MOSFET. The CNTFET device current is saturated at higher V_{ds} as channel length increases. The current level of the CNTFET decreases with channel length (when the channel length is very short) due to the energy quantization in the axial direction (as limited by optical phonon scattering).

There are two types CNTFETs: planar and coaxial CNTFETs. Whatever may be the type, these CNTFETs simply depends on the basic principles such as ballistic transport, phonon scattering and 1D density of states (DOS). Most of devices fabricated up to the date are Planar CNTFETs (Figure 2 (a)). These are popular due to their reasonable compatibility with existing manufacturing technologies and also their relative simplicity. The coaxial CNTFET is depicted in Figure 2 (b).

A lot of efforts have devoted to understand the CNTFET operation and to increase its performance. Most of the CNTFETs demonstrated to date are operates like non-conventional Schottky barrier transistors. This results a quite different scaling behavior from the conventional MOSFET. Different techniques are applied for improving the CNTFET performance. Some of these include the continuous scaling of the CNT channel, integration of thin high-κ gate dielectric insulator and utilizing the excellent source-drain metal contacts.

1.2.2 Static Random Access Memory (SRAM)

Any digital system must require some sort memory to store the data. Surveys show that throughout the world about 30% of the semiconductor business is because of the memory ICs (Sheng Lin, Yong-Bin Kim & Fabrizio Lombardi, 2008).

An SRAM is designed to fill two requirements: One need is to provide a direct interface with the CPU at speeds not attainable by DRAMs. In this the SRAM is used as cache memory, acts as an interface between the Processor and DRAMs.

Figure 2. Structures of CNTFETs: (a) planar (b) coaxial
(a) (b)
(Patil, Lin, Zhang, Wong, & Mitra, 2009)

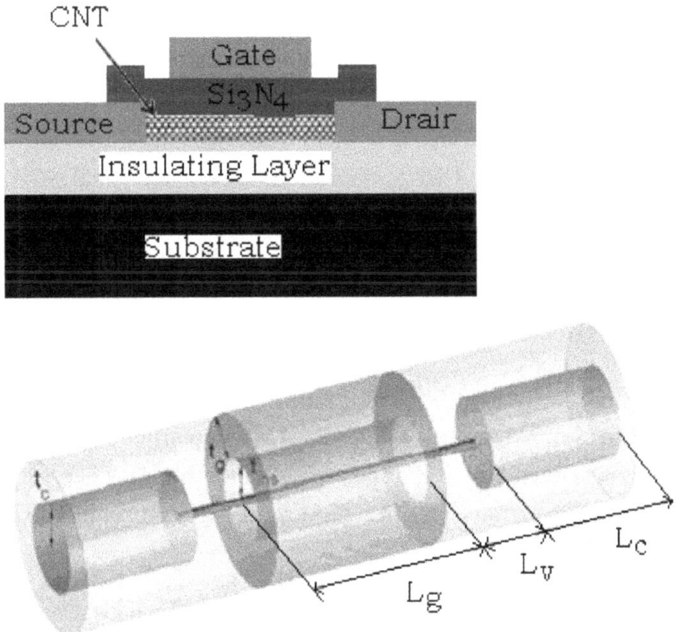

Second requirement is to replace DRAMs in systems that require very low power consumption. In this case, SRAMs are used in most portable equipment because the DRAM refresh current is several orders of magnitude more than the standby current of SRAM.

Many engineering, scientific, industrial subsystems and automotive electronics uses SRAM as its memory. Complex products such as digital cameras, cell phones etc. use SRAMs. It is also used in PCs, routers and workstations. Printers, LED and LCD screens also usually employ SRAM to hold the image to be printed or displayed. The CDRW and CDROM drives uses small amount of SRAM, to buffer track data. The cable modems and other equipment connected to computers also use SRAM.

1.2.2.1 6T CMOS SRAM Cell

The basic SRAM cell is shown in Figure 3 (Moradinasab & Fathipour, 2009). It consists of two cross-coupled inverters and two access transistors. The access transistors are connected to the wordline at their respective gate terminals, and to the bitlines at their source/drain terminals. The wordline is used to select the cell while the bitlines are used to perform read or write operations on the cell. Internally, the cell holds the stored value on one side and its complement on the other side. For reference purposes, assume that node Q holds the stored value while Qbar node holds its complement. The two complementary bitlines BL, BLbar are used to improve speed and noise rejection properties (Moradinasab & Fathipour, 2009).

Each bit in an SRAM is stored on four transistors (M1-M4) that form two cross-coupled inverters (as shown in Figure 4). This storage cell has two stable states (Q and Qbar). Two extra access transistors (M5, M6) are added to control the access to a storage bit during read and write operations. A conventional SRAM cell uses six transistors (6T) to store one bit, called 6T SRAM cell, as shown in Figure 4.

Access to the bit stored in the cell is enabled by the word line (WL) which controls the two access transistors M5 and M6 which, in turn, control whether the cell should be connected to the bit lines (BL and BLbar) or not. BL and BLbar bit lines are used to transfer data for both read and write operations. Though it is not strictly essential to have two bit lines (BL and BLbar), they are utilized to improve noise margins. The symmetric structure of SRAMs also provides differential output signals, which makes even small voltage swing on the bit lines is easily detectable.

A 6T SRAM cell is operated in three modes (de Alba Rosano & Garcia-Garcia, 2006):

1. Standby mode where the cell is idle.
2. Reading mode when the cell data has been requested.
3. Writing mode when updating the cell contents.

- **Standby State:** In this mode the cell is not accessed for either reading the data or writing the data. In this mode the word line WL is not asserted, so that

Figure 3. Basic SRAM cell

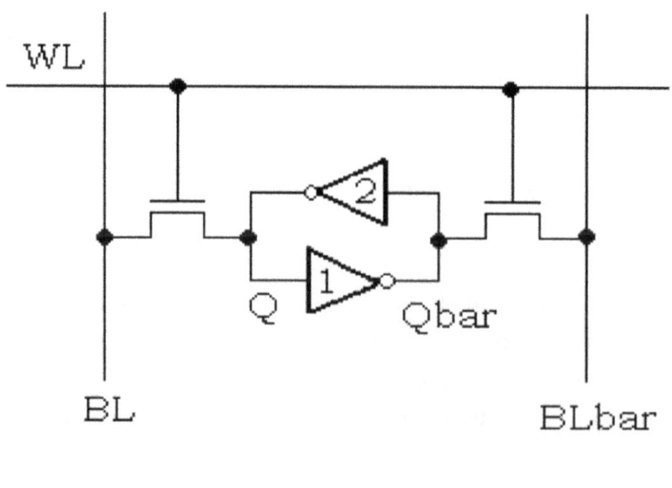

Figure 4. Basic CMOS 6T SRAM cell

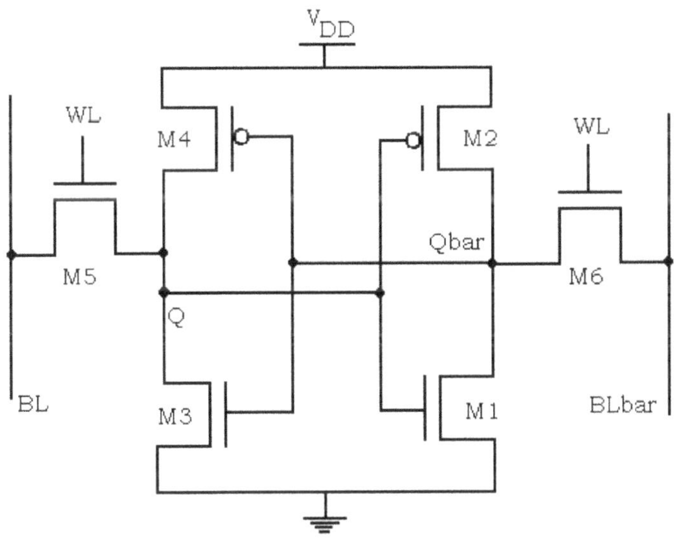

the access transistors M5 and M6 are OFF, that disconnect the cell from the bit lines. The two cross-coupled inverters formed by M1 - M4 will continue to reinforce each other as long as they are disconnected from the outside world.

- **Read Operation:** Assume that the content of the memory is a 1, stored at Q. The read cycle is started by pre-charging both the bit lines BL and BLbar to a logical '1', then asserting the word line WL, enabling both the access transistors M5 and M6 to ON. In the second the values stored in Q and Qbar are transferred to the bit lines by leaving BL at its pre-charged value of '1' and discharging BLbar through M1 and M6 to a logical '0'. On the BL side, the transistors M4 and M5 pull the bit line toward V_{DD}, a logical 1. If the content of the memory assumed to be a '0', the opposite thing is going to happen that is BLbar is pulled toward 1 and BL is pulled toward 0.
- **Write Operation:** To write a value into an SRAM cell, the new value and its complement are applied on the bit lines (BL and BLbar). As an example, to write a '0' into the SRAM cell, a '0' is applied to BL and '1' to BLbar. A '1' is written into the cell by applying reverse combination of the values for the bit lines, i.e., applying '1' to BL and '0' to BLbar. After that the word line (WL) is raised. Then the new value will overwrite the old value. To happen this writing possible the write circuitry transistors are to be designed much stronger than the DRAM cell transistors, so that they can easily supersede the previous stored bit state of the cross-coupled inverters. Careful sizing of the transistors in an SRAM cell is needed to ensure proper operation.

1.2.2.2 6T CNTFET SRAM Cell

The design of 6T SRAM cell using CNTFETs is shown in Figure 5.

This uses six numbers of CNTFETs M1-M6 to store one data bit. This CNTFET cell is also operates in standby, read and write modes. For the successful write operation, internal capacitor at the node (Q or Qbar) must discharge its voltage below the V_{TH}, so that the node flips the state.

1.2.2.3 6T CNTFET SRAM Cell Leakage Paths

The sub-threshold leakage current in an 6T CNTFET SRAM cell is usually categorized into two kinds as shown in Figure 6: (i) cell leakage current that flows from V_{DD} to GND, internal to the cell and (ii) bitline leakage current that flows from BL (or BLbar) to GND.

Although a 6T CNTFET SRAM cell has two bitline leakage paths, the BL leakage current and BLbar leakage current differ according to the value stored in the SRAM cell. If a 6T CNTFET SRAM cell holds '1', the BL leakage current passing through M5 and M3 is effectively suppressed due to two reasons. First, after precharging BL and BLbar both to '1,' the source voltage and the drain voltage of M5 are the same, and thus potentially no current flows through M5. Second, two stacked and

Figure 5. Basic 6T CNTFET SRAM cell

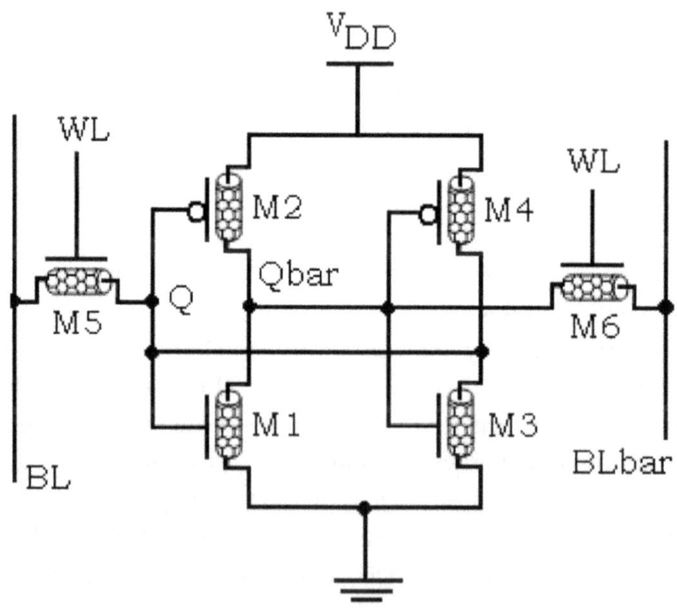

turned off transistors (M5 and M3) induce the stack effect (Johnson, Somasekhar, Chiou, & Roy, 2002).

Meanwhile, for this case where the 6T CNTFET SRAM cell holds value '1,' a large BLbar leakage current flows passing through M6 and M1. If, on the other hand, if the 6T CNTFET SRAM cell holds '0,' a large bitline BL leakage current flows while BLbar leakage current is suppressed.

Figure 7 and Figure 8 show various leakage currents flowing through 6T CNTFET SRAM cell for the assumed storage bit of '0' and '1' respectively (Yu, Chen, Nan, Wang & Choi, 2011). I_{sub} and I_{gate} represent sub-threshold and gate leakage currents respectively.

Notice that if BL and BLbar lines are pre-charged to V_{DD}, the drain-to-source voltage of M5 and M6 becomes zero and according to (1) and (2) the sub-threshold currents through these transistors, i.e., I_{sub5} and I_{sub6}, will be zero, for Q='1' and Q='0' respectively.

Therefore, the total leakage current of 6T CNTFET SRAM cell is:

$$I_{Total} = I_{sub2} + I_{sub3} + I_{sub5} + I_{gate1} \text{ if Q='0'} \text{ ------} \quad (1)$$

$$I_{Total} = I_{sub1} + I_{sub4} + I_{sub6} + I_{gate3} \text{ if Q='1'} \text{ ------} \quad (2)$$

Figure 6. Leakage paths in conventional 6T CNTFET SRAM cell

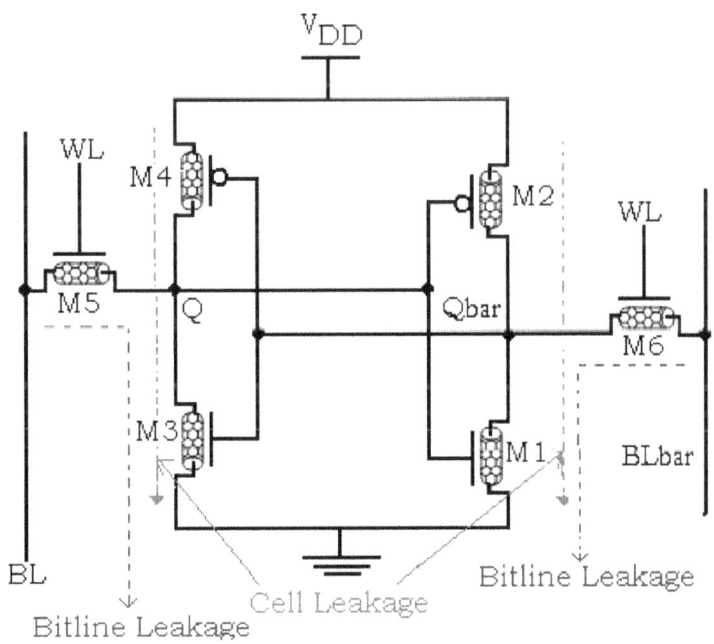

Figure 7. Major leakage currents in a 6T CNTFET SRAM cell with Q='0'

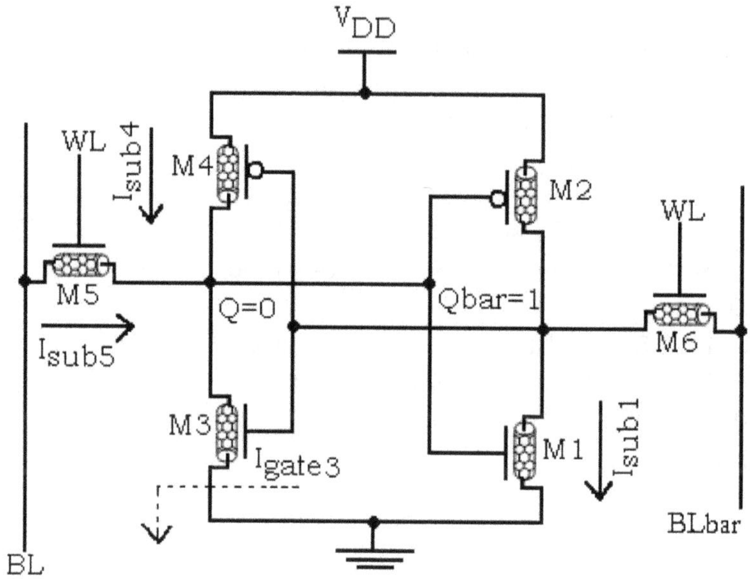

Figure 8. Major leakage currents in a 6T CNTFET SRAM cell with Q='1'

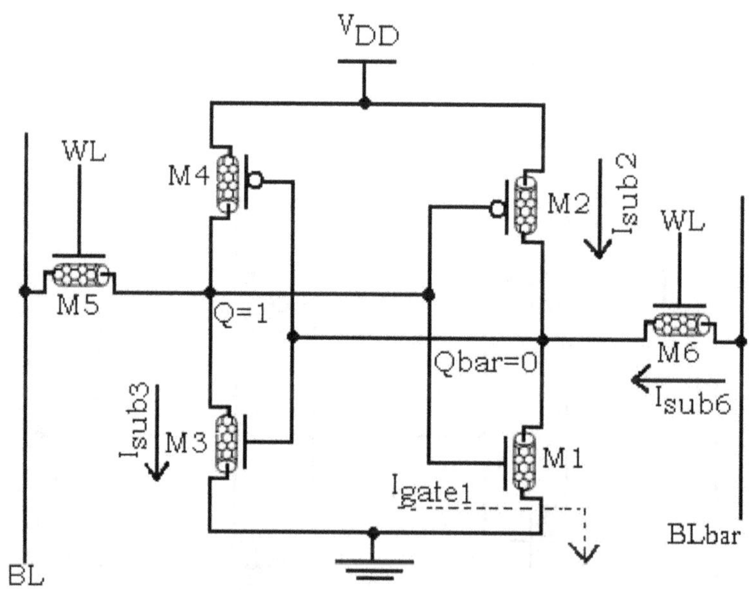

1.3 Design and Analysis of CNTFET-Based 6t SRAM Cells

This section describes, design of 6T CNTFET SRAM cell using different circuit level leakage power reduction techniques. It also describes assessment and comparison of this 6T CNTFET SRAM ccll with CMOS-bascd 6T SRAM ccll.

1.3.1 6T CNTFET SRAM Cell Design

The conventional six-transistor (6T) SRAM memory cell configuration used as the core storage element in most cache designs for low-power embedded applications. With today's aggressive scaling, substantial problems such as power consumption and stability have already been encountered when the 6T SRAM cell configuration is utilized in CMOS at nanoscale ranges. In this section, the 6T SRAM cell is designed using CNTFETs as shown in Figure 9 and its performance is assessed to compare power dissipation and stability with standard CMOS 6T SRAM cell design.

Read Operation

To read a value from a 6T CNTFET SRAM cell, the bitlines (BL and BLbar) are first pre-charged to logic 1. Then the wordline signal WL is set to high, to make the CNTFET access transistors M5 and M6 are ON. Because one of the internal storage nodes (Q or Qbar) is low, one of the bit lines (BL and BLbar) starts discharging. A

Figure 9. Basic 6T CNTFET SRAM cell

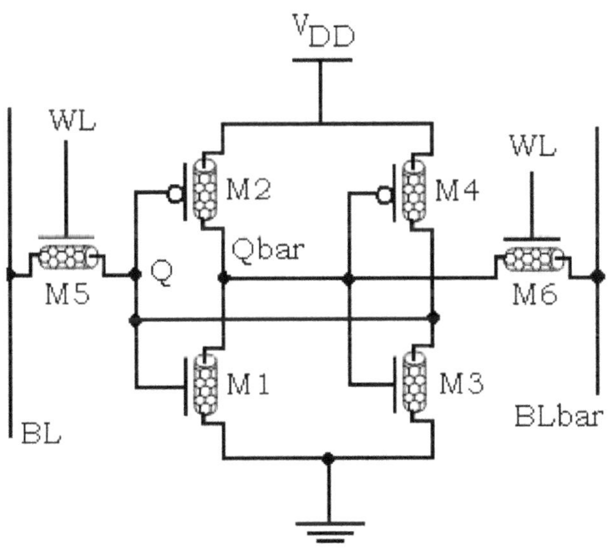

sense amplifier which is connected to the bit lines (BL and BLbar) senses which of the bit lines is discharging and reads the stored value. Careful sizing of the CNTFET transistors is required for proper operation of the 6T CNTFET SRAM cell, otherwise a read-upset problem is going to present, and this problem may change the data stored in the 6T CNTFET SRAM cell. The read-upset problem can be described as follows. Assume that the 6T CNTFET SRAM cell is currently storing "1" so that Q is "1" and Qbar is "0". When WL is high, the CNTFET access transistors (M5 and M6) are ON. The voltage at node Qbar will rise as its gets charged to BLbar through M6. An appropriate sizing ratio between M1 and M6 is required to limit the voltage at node Qbar to be lower than V_{TH} such that the stored logic value does not change during the read operation. In the conventional CMOS 6T SRAM cell design, the M1/M6 ratio should be greater than 1.28 for this requirement (Wang, Yu & Choi, 2011). For this 6T CNTFET SRAM cell design, simulations have been performed to establish the sizing ratio of M1 and M6. The right ratio of M1/M6 is found by running simulations for various M1/M6 ratios vs. the voltage rise at Qbar node. The transistor size ratio of the two CNTFETs is measured as the number of tubes in the two CNTFETs unlike MOSFET in which it is width to length ratio of the transistor. The threshold voltage of the (19, 0) chirality CNTFET is $|0.289|$V. After several iterations the M1/M6 ratio of this 6T CNTFET SRAM cell found that it should be greater than 1.4 to keep the voltage of Qbar below V_{TH}. However, in this design the M1/M6 ratio of 1.5 used for fair comparisons.

Write Operation

During the write operation, the wordline WL is high to allow the data on bitlines BL and BLbar to be written into the 6T CNTFET SRAM cell. For a successful writing of data into a 6T CNTFET SRAM cell, the pull up PCNTFEts (M2 and M4) should not be too strong. Assume that the 6T CNTFET SRAM cell is storing "0" and it is required to write a new data "1" into the cell. For proper writing, the pass NCNTFETs (M5 and M6) must be significantly more conductive than the pull up PCNTFETs (M2 and M4). In the traditional CMOS design, the M2/M6 ratio should not be greater than 1.6 (Wang, Yu & Choi, 2011). For this 6T CNTFET SRAM cell design, simulations have been performed to establish the size ratio between M2 and M6. Simulations are run for various ratios. Any M2/M6 ratio of less than 0.5 can pull node Q below 0.289 V, which is the threshold voltage of a CNTFET with (19, 0) chirality nanotubes. Therefore, for this 6T CNTFET SRAM cell design, the transistor size ratios among the pull-up CNTFETs, the pull down CNTFETs, and the access transistor CNTFETs are found to be M2/M6 = 0.5 and M1/M6 = 1.5.

In this 6T CNTFET SRAM cell design, two tubes are used for PCNTFET driver transistors (M2 and M4), four tubes are used for NCNTFET access transistors (M5 and M6) and six tubes for NCNTFET driver transistors (M1 and M3).

1.3.1.1 Simulation Results and Discussions

Since the V_{TH} of CNTFET can be controlled by adjusting tube's diameter, the design of CNTFET-based circuits with different V_{TH}'s is possible because CNTs can be grown with different diameters (Wang, Yu & Choi, 2011). In this section, NCNTFETs and PCNTFETs use CNTs that have different chirality vectors for the best performance. However, all NCNTFETs use CNTs with the same chirality vector and all PCNTFETs use the same chirality vector as well.

The difference in chirality between NCNTFETs and PCNTFETs must also take into account for the performance of the SRAM memory cell. As for CMOS 6T SRAM cell, the V_{TH}'s of the pull-up PMOSs (M2 and M4) have a close relationship with the SNM of the 6T SRAM cell. The SNM is defined as the maximum value of DC noise voltage that can be tolerated by the SRAM cell without changing the stored bit. The SNM is commonly used as a metric for static stability of a SRAM cell.

To investigate the SNM of the proposed 6T CNTFET SRAM, extensive simulations have been performed on 6T CNTFET SRAM cell. HSPICE simulations are performed using 32nm Stanford CNTFET model (http://nano.stanford.edu/model.php?id=23) for 6T CNTFET SRAM cell and Berkeley Predictive Technology (http://wwwdevice.eecs.berkeley.edu/~ptm/) 32nm model is used for simulating CMOS 6T SRAM cell. For the CMOS 6T SRAM cell configuration, the worst-case stability condition occurs when the cell is accessed for read operation, i.e. the read SNM is lower than the hold SNM. When the chirality vector of the PCNTFETs changes from (10, 0) to (22, 0), the SNM of the 6T CNTFET SRAM is increased. The values of the read SNM of the CNTFET SRAM cell change little as the chirality vectors of the NCNTFETs changes. Therefore, for best stability, the chirality vector of the PCNTFETs must be adjusted. In this design chirality vector of PCNTFET is varied from (10, 0) to (22, 0) by keeping chirality vector of NCNTFET constant at (19, 0).

The stability of the 6T CNTFET SRAM cell can be increased by decreasing the absolute value of the V_{TH} of the pull-up PCNTFET by controlling its chirality vector (n, m). However, there is a conflicting constraint between performance and stability. At a better ability to hold data, it is also harder to write new data into the 6T CNTFET SRAM cell, i.e. it takes more time to write new data. To find the optimum chirality for both PCNTFETs and NCNTFETs, both the SNM and the write time must be considered. For a highly stable and low delay design, a high SNM and a fast write time are desired. Therefore, the SNM is divided by the write time to find the best V_{TH} of the PCNTFET for both high SNM and fast write time. The (19, 0)

6T CNTFET SRAM cell is selected for best overall performance. Table 1 shows the comparison of 6T CNTFET SRAM cell with CMOS 6T SRAM cell in terms of leakage power, dynamic power, read delay, write delay and SNM. The leakage power in 6T CNTFET SRAM cell is reduced by 93.33% and SNM indicates stability increased by 57.25% compared to CMOS 6T SRAM cell.

For SRAMs Static Noise Margin (SNM) is one of the important performance metrics as it indicates the stability of stored data w.r.t noise or some other disturbance. For best stability, the chirality vector of the PCNTFETs must be adjusted by keeping the chirality of NCNTFET at constant.

1.3.2 Sleep Transistor 6T CNTFET SRAM Cell Design

In earlier micrometer technologies, dynamic power dominates the total IC power, but in present nanometer technologies leakage power dominates the total power. This method reduces leakage power by dynamically disconnecting the power supply VDD to the 6T CNTFET SRAM cell, when it is in standby mode.

1.3.2.1 *Sleep Transistor Technique Principles*

The most natural way of lowering the leakage power dissipation of a VLSI circuit in the standby mode is to turn OFF its power supply V_{DD}. This can be done by using one PCNTFET and one NCNTFET sleep transistors in series with the other transistors of each logic block to create a virtual ground and a virtual V_{DD} as shown in Figure 10. These sleep transistors are controlled by using 'Sleep' and 'Sleepbar' control signals.

In the active mode, the sleep transistors are ON, thereby the circuit functions as usual. In the standby mode, the sleep transistors are turned OFF, which disconnects the circuit from the power supplies rails V_{DD} and GND. To lower the leakage power, the V_{TH} of the sleep transistors must be large. Otherwise, the sleep transistors itself have a high leakage current, which makes the sleep transistor technique less effective.

Table 1. Simulation results of 6T CNTFET SRAM cell and CMOS 6T SRAM cell

Sl. No.	Parameter	6T CNTFET SRAM Cell	CMOS 6T SRAM Cell
1	Leakage Power (W)	6.85e-09	1.13e-07
2	Dynamic Power (W)	4.63e-07	2.01 e-06
3	Read Delay (s)	7.43e-12	25.1 p
4	Write Delay (s)	6.5e-12	35.3p
5	SNM (mV)	195	124

Additional savings may be achieved if the widths of the NCNTFET and PCNTFET sleep transistors are smaller than the combined width of the CNTFETs in the pull-down and pull-up networks respectively.

Transistors with a low V_{TH} are used to implement the logic, while transistors with high-V_{TH} are used as sleep transistors as shown in Figure 11.

To guarantee the proper functionality of the circuit, the sleep CNTFETs have to be sized carefully to decrease their ON state voltage drop. The voltage drops on these sleep CNTFETs decreases the effective V_{DD} of the circuit. This problem can be solved by using a large size CNTFET sleep transistors. On the other hand, using a large size CNTFET sleep transistor increases the area overhead and also the dynamic power consumed for turning the CNTFETs ON and OFF. Because of this dynamic power consumption, it is not possible to save power using this sleep transistor technique for short periods of standby mode. There should be a minimum duration of the standby time below which power saving is impossible. As the size

Figure 10. Basic principle of sleep transistor technique

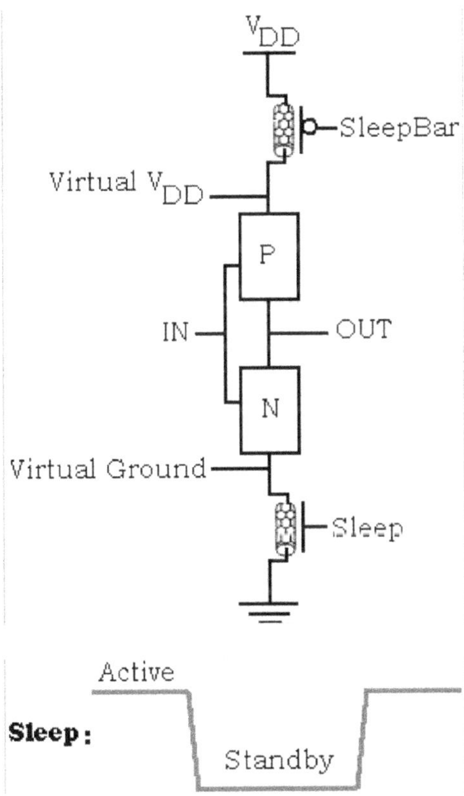

of the sleep CNTFETs increases, this minimum duration of standby mode also increases. Sleep transistor technique is a very effective method for decreasing the leakage power (Kim, Nan & Choi, 2010).

1.3.2.2 *Design of Sleep Transistor 6T CNTFET SRAM Cell*

Sub-threshold leakage current and leakage power dissipation increase exponentially with decreasing V_{TH}. To reduce leakage power dissipation in a 6T CNTFET SRAM cell from limiting aggressive V_{TH} scaling, an existing sleep transistor technique is applied in this section. This sleep transistor technique enables a 6T CNTFET SRAM cell to turn OFF the power supply and eliminate virtually all the leakage power dissipation in the 6T CNTFET SRAM cell. The key idea of this technique is to introduce an extra two CNTFETs, one is in between the power supply V_{DD} and the 6T CNTFET SRAM cell; another one is in between the 6T CNTFET SRAM cell and the ground GND. These two extra CNTFETs are turned ON when the 6T CNTFET SRAM cell is in active mode, that is when the cell is accessed for reading/ writing and turned OFF when 6T CNTFET SRAM cell is in idle or standby or hold mode, that is when it is not accessed for reading/writing. Sleep transistor technique maintains the performance advantages of lower power supply and threshold voltages by reducing leakage power dissipation.

The 6T CNTFET SRAM cell with existing sleep transistor technique is shown in Figure 12. PCNTFET sleep transistor M8 is connected between V_{DD} and the 6T CNTFET SRAM cell PCNTFET transistors and NCNTFET sleep transistor M7 is

Figure 11. Conventional sleep transistor scheme

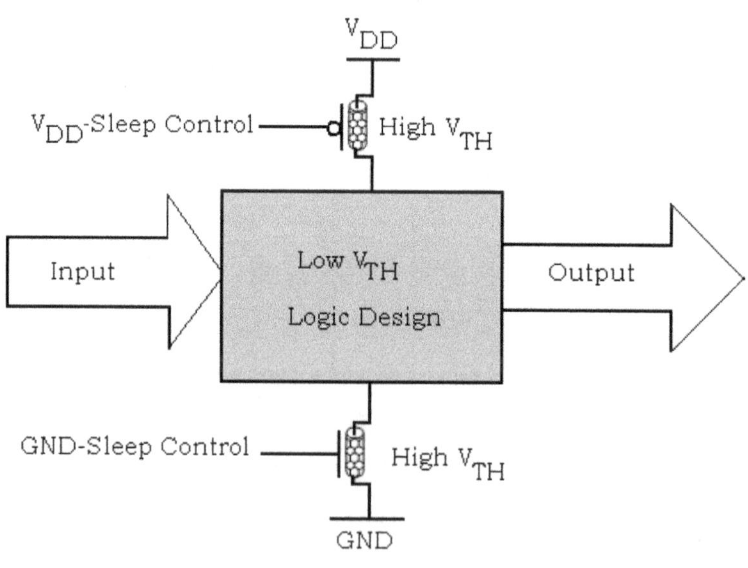

connected between GND and the 6T CNTFET SRAM cell NCNTFET transistors. These sleep CNTFETs M7 and M8 are turned ON for the cell to be in active mode and turned OFF for the cell to be in standby mode. These two sleep CNTFETs M7 and M8 are controlled by a signal called 'Sleep'. This control signal is '1' in active mode and '0' in standby mode, so that the sleep CNTFETs are ON in active mode and OFF during standby mode.

This CNTFET based SRAM cell is in active mode when it is accessed for reading/writing and it is in standby mode when it is not accessed. The accessing or not accessing of sleep transistor 6T CNTFET SRAM cell is controlled by using a control signal called word line (WL). The control signal 'WL' is set to '1' to access the cell and it is set to '0' when it is not accessed. When a control signal 'WL' is '0', control signals 'Sleep' and 'SleepBar' are set to '0' and '1' respectively so that sleep CNTFETs M7 and M8 are OFF, there by leakage currents are reduced and hence leakage power.

Figure 12. Design of sleep transistor 6T CNTFET SRAM cell

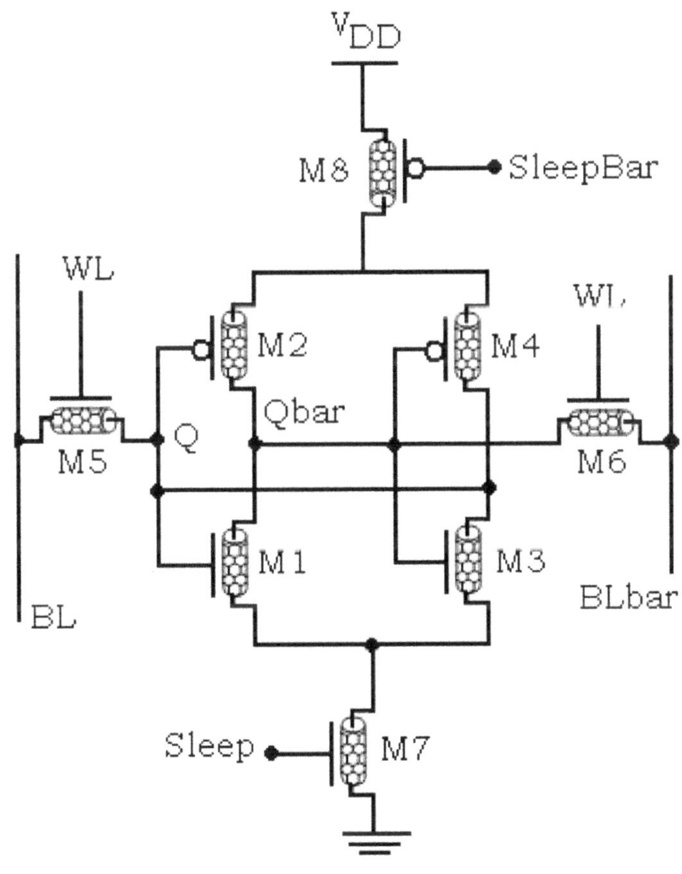

These Sleep Transistors M7 and M8 can be shared among multiple 6T CNTFET SRAM cells to minimize the area overhead. To reduce the impact on 6T CNTFET SRAM cell speed and to ensure stability of the 6T CNTFET SRAM cell, the sleep CNTFETs must be carefully sized with respect to the 6T CNTFET SRAM cell transistors when they are gating. While these sleep CNTFETs must be made large enough to sink the current flowing through the 6T CNTFET SRAM cell during a read/write operation in the active mode, too large a sleep CNTFET, may reduce the stacking effect, thereby diminishing the power savings. Moreover, large CNTFET also increase the area overhead.

In practice only one sleep transistor (either NCNTFET or PCNTFET) is sufficient, because of their lower on-resistance, NCNTFET transistors are usually used. But to get large leakage reduction two sleep CNTFETs are used, but at the cost of increase in area.

1.3.2.3 *Simulation Results and Discussions*

Synopsys HSPICE circuit simulator is used for simulating sleep transistor 6T CNTFET SRAM cell. Simulations performed with Stanford CNTFET model at 32nm feature size with supply voltage V_{DD} of 0.9V (http://nano.stanford.edu/model.php?id=23). The HSPICE CScope is used for displaying simulated waveforms. Simulated results of 6T CNTFET SRAM cell with and without sleep transistors are summarized in Table 2 for comparison. The leakage power in sleep transistor 6T CNTFET SRAM cell is reduced by 95.25% compared to the base case 6T CNTFET SRAM cell but at the cost of 6.15% write delay increase and 8.33% area increase.

1.3.3 Forced Stack 6T CNTFET SRAM Cell Design

It has been observed that the small leakage current is flowing through stacking of two OFF transistors compared to leakage current flowing through the one OFF transistor.

1.3.3.1 *Basics of Forced Stack Technique*

Leakage current through series-connected transistors or transistor stacks, with more than one transistor OFF, is at least an order of magnitude smaller than that through single OFF transistor. This so called stack-effect can be utilized for reducing leakage power in VLSI circuits. Stacking of transistors can reduce sub-threshold leakage (Johnson, Somasekhar, Chiou, & Roy, 2002), so it is called stacked effect. The effect of stacking on reducing leakage can be understood from the Figure 13. Leakage reduction by stack effect can be exploited by converting a single transistor into a stack of two-transistors. The widths of these transistors can be half of the original

Table 2. Simulation results of 6T CNTFET SRAM cell

Sl. No.	Parameter	Base Case	With Sleep Transistor Technique
1	Leakage Power (W)	6.85e-09	0.325e-09
2	Dynamic Power (W)	4.63e-07	4.635e-07
3	Read Delay (s)	7.43e-12	7.81e-12
4	Write Delay (s)	6.5e-12	6.9e-12
5	SNM (mV)	195	32.23
6	Area (nm^2)	15360	16640

transistor size or even other combinations can also be chosen to preserve the same input capacitance load as the original single transistor.

Figure 13 (a) shows single CNTFET and Figure 13 (b) shows connecting two CNTFETs as stack. In Figure 13 (b), if the input is '0', then both CNTFETs M1 and M2 are turned OFF. Here V_A is the intermediate node voltage between M1 and M2 CNTFETs. CNTFET M1 has its internal resistance. Due to this resistance V_A is greater than the ground potential. This positive V_A results in a negative gate-source (V_{gs}) for the M2 CNTFET and the negative base-source voltage (V_{bs}) for M2. Here M2 also has a reduced drain-source voltage (V_{ds}), which lowers the drain induced barrier lowering (DIBL) effect. These three effects together reduce the leakage power. The Figure 14 also illustrates the concept of stacking effect. This figure shows the leakage current difference between a single OFF transistor and a stack of two OFF transistors.

Stacking of series connected OFF CNTFETs can be used for the reduction of the sub threshold leakage currents. This effect is called the forced stacking [24], yields the stacking effect by inserting extra CNTFET for every input of the circuit in both PCNTFET and NCNTFET networks. So in the forced stacking two CNTFETs are always OFF for every OFF input of the circuit, which reduces the leakage current and hence leakage power.

1.3.3.2 *Forced Stack 6T CNTFET SRAM Cell*

The 6T CNTFET SRAM cell design by applying forced stack technique to reduce leakage power is shown in Figure 15. Two pairs of CNTFET stack transistors (M7, M9 and M8, M10) are used in the 6T CNTFET SRAM cell. One in each pair is activated during idle mode based upon the value of the bit stored in the cell. This disconnects the OFF CNTFETs from supply while retaining supply to the ON CNTFETs. The effect of stacking the CNTFETs results in the reduction of sub-

Figure 13. (a) Single transistor (b) Forced stack transistor

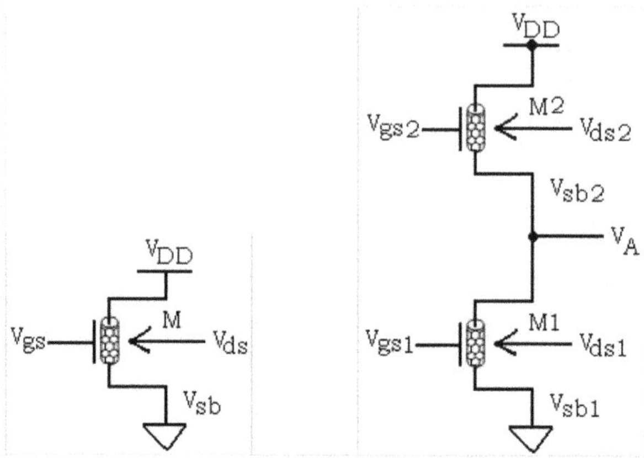

Figure 14. Leakage current difference between a single OFF device and a stack of two OFF devices

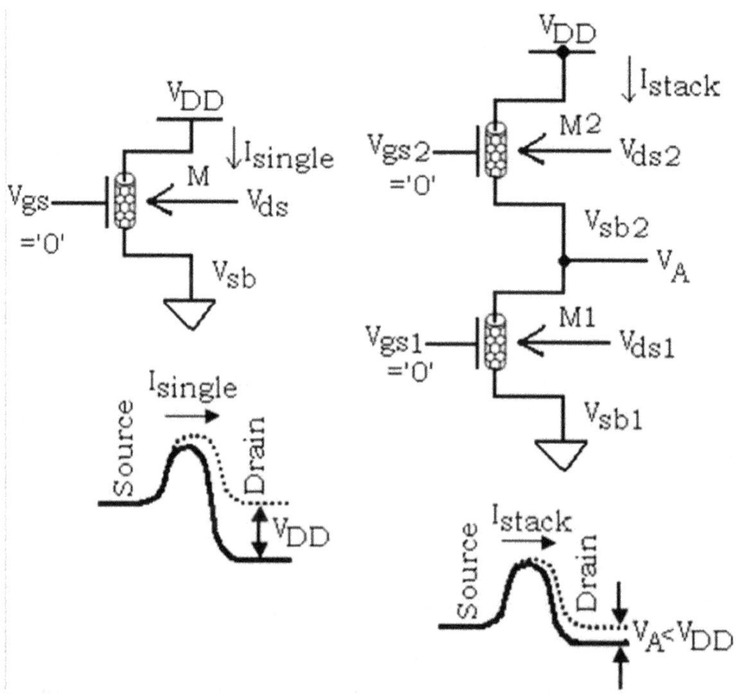

threshold leakage current when two or more CNTFETs are turned OFF together (Johnson, Somasekhar, Chiou, & Roy, 2002).

In the conventional 6T CNTFET SRAM cell there are only 6 CNTFETs (M1-M6). But in this forced stack 6T CNTFET SRAM cell two extra pull up PCNTFETs (M8 and M10) and two extra pull down NCNTFETs (M7 and M9) are added in addition to original six CNTFETs. M1, M2, M3, M4, M5 and M6 are original six CNTFETs. All the CNTFETs share the same input in the forced stack 6T CNTFET SRAM cell. This forced stack technique is a state saving technique. That means when the circuit is in standby mode it retains the stored bit value. The main drawback of this forced stack 6T CNTFET SRAM cell is that it cannot use the high V_{TH} for CNTFETs. Because if it use the high V_{TH} CNTFETs than there is a dramatic increase of delay.

In CNTFETs the number of tubes used for each transistor decides size of the transistor. In this forced stack 6T CNTFET SRAM cell one tube is used for all PCNTFET driver and stack transistors (M2, M4, M8, M10), three tubes are used for NCNTFET driver and stack transistors (M1, M3, M7, M9) and four tubes are used for NCNTFET access transistors (M5, M6). The forced stack technique can achieve leakage power saving while retaining the logic state.

Figure 15. Forced stack 6T CNTFET SRAM cell

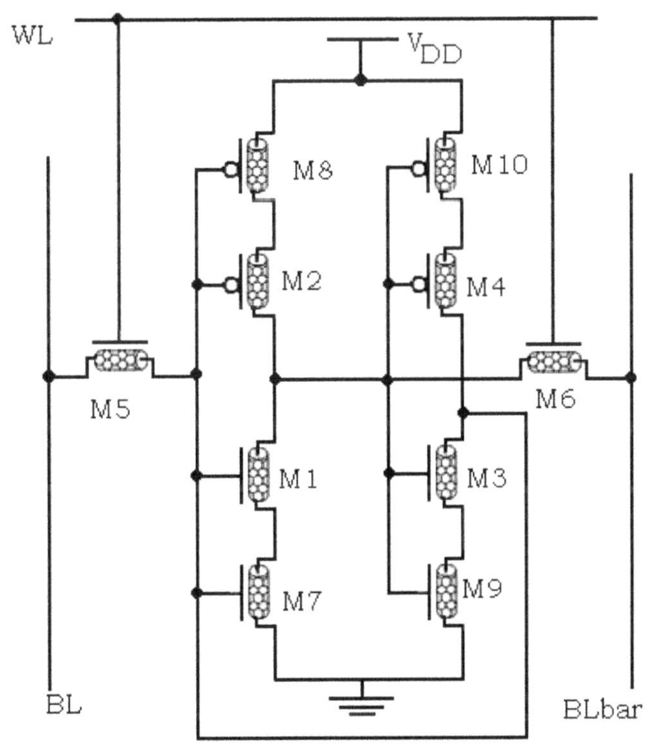

1.3.3.3 *Simulation Results and Discussions*

The 6T CNTFET SRAM cell is designed by applying existing forced stack technique and it is simulated by using the Synopsys HSPICE circuit simulator. Simulations are performed with Stanford CNTFET model at 32nm feature size with supply voltage V_{DD} of 0.9V (http://nano.stanford.edu/model.php?id=23).

The HSPICE circuit simulator is used for measuring leakage currents. This technique is compared to the base case and sleep transistor technique in terms of leakage power, dynamic power, read delay, write delay, SNM and area and these simulation results are tabulated in Table 3. This forced stack 6T CNTFET SRAM cell reducing leakage power by 45.54% compared to base case at the cost of dramatic increase of 60.43% in read delay. The advantage of this forced stack 6T CNTFET SRAM cell compared to sleep transistor 6T CNTFET SRAM cell is that it retains the state while the sleep transistor technique cannot.

1.3.4 Data-Retention Sleep Transistor Technique

Earlier a lot of research is carried out to reduce leakage power of SRAM. Out of these, sleep transistor leakage reduction technique is the effective method. The key idea of the sleep transistor cell is to cut-off the cell from the V_{DD} and ground when it is not working. But the drawback of this disconnection of the cell from supply and ground is that there is chance for the virtual ground node may get charged to V_{DD} or the virtual supply node may get discharged to '0', due of this the stored bit in this cell may get destroyed, so this sleep transistor cell cannot hold stored data during standby mode.

Table 3. Simulation results of 6T CNTFET SRAM cell

Sl. No.	Parameter	With Forced Stack Technique	With Sleep Transistor Technique	Base Case
1	Leakage Power (W)	3.73e-09	0.325e-09	6.85e-09
2	Dynamic Power (W)	4.631e-07	4.635e-07	4.63e-07
3	Read Delay (s)	11.92e-12	7.81e-12	7.43e-12
4	Write Delay (s)	10.3e-12	6.9e-12	6.5e-12
5	SNM (mV)	193	32.23	195
6	Area (nm^2)	15360	16640	15360

1.3.4.1 *Data Retention Voltage (DRV)*

Data Retention Voltage (V_{DRV}) is the minimum power supply voltage required to retain the stored data in the standby mode. The nodes (Q or Qbar) of the 6T CNTFET SRAM cell are used for storing bit of value either 0 or 1. To find the DRV of 6T CNTFET SRAM cell, decrease the V_{DD} until the cell flips the state. If Q=1 and Qbar=0 then they are going to change to values of Q=0 and Qbar=1, if V_{DD} is decreases below V_{DRV}. When V_{DD} scales down to DRV, the Voltage Transfer Curves (VTC) of the internal inverters degrade to such a level that SNM of the 6T CNTFET SRAM cell reduces to zero. DRV should be greater than V_{TH}. After several iterations the DRV of this cell is found to be 0.3V, as shown in Figure16.

1.3.4.4 *Data-Retention Sleep Transistor 6T CNTFET SRAM Cell*

This section going to show that in the data-retention sleep transistor technique, maximum leakage reduction is achieved only when both NCNTFET and PCNTFET sleep transistors are used. The data-retention 6T CNTFET SRAM cell using both NCNTFET and PCNTFET sleep transistors is shown in Figure 17.

For the data-retention sleep transistor 6T CNTFET SRAM cell to be accessed for a read or write operation, the signals WL and 'Sleep' become '1', which causes the voltage of the virtual supply and virtual ground nodes to become V_{DD} and GND,

Figure 16. Finding DRV of a 6T CNTFET SRAM cell

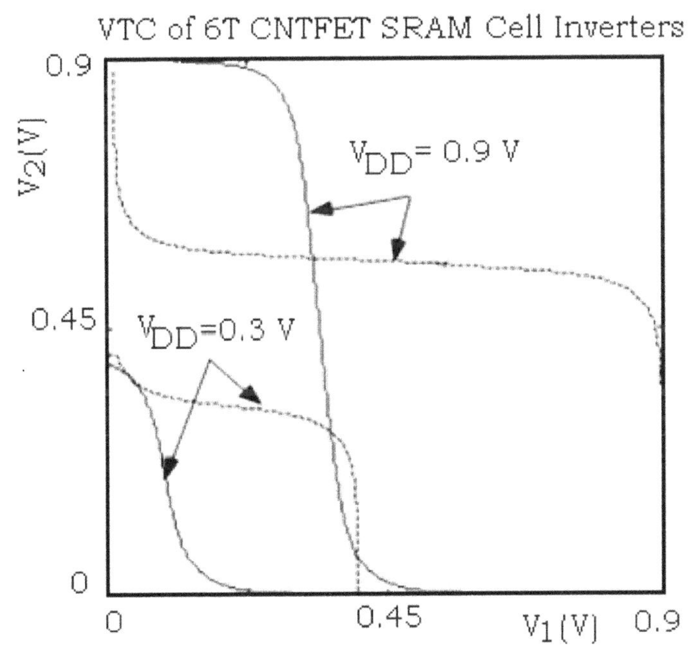

respectively. As soon as the operation is completed, the signals WL and 'Sleep' become '0' and the data-retention 6T CNTFET SRAM cell enters into the standby mode. In this mode, the sleep CNTFETs M7 and M8 are OFF and the strapping CNTFETs M9 and M10 turns ON. This results in the voltage of virtual ground and virtual supply become V_{GND} and V_{VDD}, respectively.

This data-retention 6T CNTFET SRAM cell reduces leakage power due to source body biasing of the pull-up PCNTFETs (M2, M4), pull-down NCNTFETs (M1, M3), and access NCNTFETs (M5, M6). Like single NCNTFET sleep transistor and single PCNTFET sleep transistor 6T CNTFET SRAM cells, the leakage power in this 6T CNTFET SRAM cell is reduced if the voltage difference (i.e., $\Delta V = V_{VDD} - V_{GND}$) between the two power supply rails of the cell is small; however, it also makes the cell more susceptible to noise there by losses the stored bit. This data-retention sleep transistor 6T CNTFET SRAM cell using both NCNTFET and PCNTFET sleep transistors is compared to single NCNTFET data-retention sleep transistor 6T

Figure 17. Data-retention sleep transistor 6T CNTFET SRAM cell

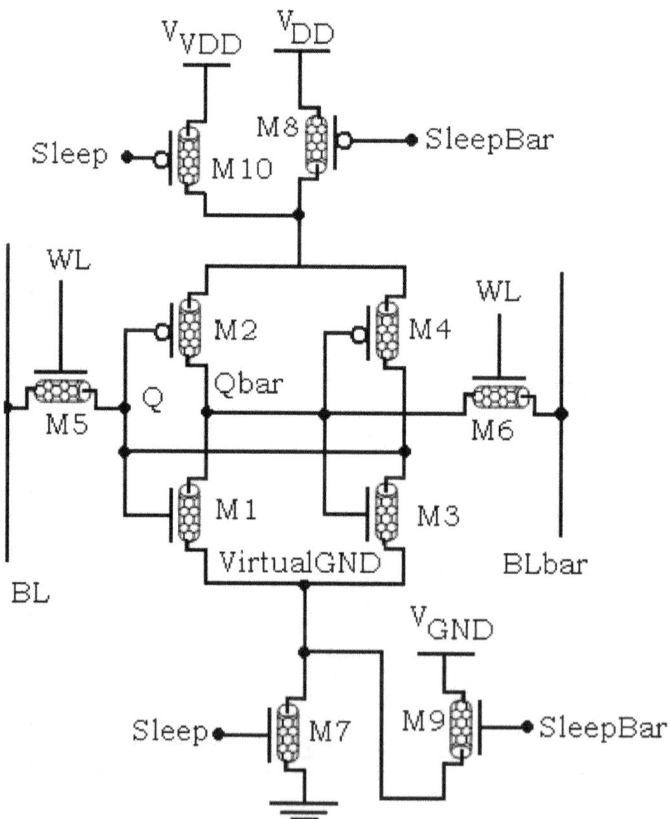

CNTFET SRAM cell in terms of leakage power, write delay, SNM and area. For fair comparison they have compared when they have equal ΔV in the standby mode.

Notice that if ΔV is too low, all six transistors will work in the sub-threshold region, but as long as ΔV is greater than the DRV of the 6T CNTFET SRAM cell, the data is retained.

Consider Figure 17 and assume a fixed $\Delta V = V_{VDD} - V_{GND}$. If both V_{VDD} and V_{GND} are increased such that their difference remains fixed, I_{sub3}, I_{sub5}, and I_{sub6} are lowered because of the stacking and the body effect, yet I_{sub2} is increased because of the lower bulk-to-source voltage, which results in a more positive V_{TH} for PCNTFET transistor M2. At the same time, I_{gate1} remains constant because the voltage potential across its gate oxide is constant. On the other hand, if both V_{VDD} and V_{GND} are decreased such that their difference remains fixed, I_{sub2} is decreased because of the larger bulk-to-source voltage, while I_{sub3}, I_{sub5}, and I_{sub6} are increased because of the lower bulk-to-source voltage. This shows that when ΔV is fixed, there are optimum values for V_{VDD} and V_{GND} for which the leakage power dissipation of the 6T CNTFET SRAM cell is at the minimum.

The NCNTFET sleep transistor (M7) is in the read path should be sufficiently large to result in low delay penalty. Since the PCNTFET sleep transistor (M8) is in the write path and the write delay of 6T CNTFET SRAM cell is usually lower than the read delay, the PCNTFET sleep transistor (M8) can be made smaller than the NCNTFET sleep transistor (M7). On the other hand, the strapping transistors M9 and M10 are turned ON during the standby mode and because entering the standby mode is not a time-critical step, these strapping transistors can be very small.

1.3.4.5 *Simulation Results and Discussions*

The 6T CNTFET SRAM cell is designed at 32nm technology by applying this data-retention sleep transistor technique. This circuit is simulated in HSPICE using Stanford CNTFET model (http://nano.stanford.edu/model.php?id=23).

In data retention sleep transistor 6T CNTFET SRAM cell having a smaller voltage difference between the two power rails of the cell, i.e., $\Delta V = VV_{DD}$-VV_{GND}, in the standby mode results in lower leakage; however, it also makes the cell more susceptible to noise in the standby mode. Notice that if ΔV is too low, all six transistors will work in the sub-threshold region, but as long as ΔV is greater than the DRV of the cell, the stored data is retained. If $\Delta V = VV_{DD}$-VV_{GND} is fixed, leakage currents through OFF transistors are lowered because of the stacking and the body effect. Analysis shows that when ΔV is fixed, there are optimum values for VV_{DD} and VV_{GND} for which the leakage power dissipation of the cell is at the minimum. In these simulations both 'BL' and 'BLbar' are assumed to be pre-charged to V_{DD}.

This data-retention sleep transistor 6T CNTFET SRAM cell is compared to other existing leakage power reduction techniques based cells, called sleep transistor 6T CNTFET SRAM cell, forced stack 6T CNTFET SRAM cell and also with base case in terms of leakage power, dynamic power, read delay, write delay, SNM and area. These simulation results are tabulated in Table 4.

1.3.5 Stacked Sleep 6T CNTFET SRAM Cell

1.3.5.1 *Structure of Stacked Sleep Technique*

The structure of this stacked sleep technique combines the structures of the two popular leakage reduction techniques, called sleep transistor and forced stack. In this stacked sleep technique, sleep transistors are getting stacked, that's why the name stacked sleep. This technique structure merges structures of sleep transistor and forced stack. The stacked sleep inverter is shown in Figure 18.

1.3.5.2 *Operation of Stacked Sleep Technique*

The operation of this stacked sleep technique is explained by taking inverter as example and considering its active and standby modes. The sleep transistors used in this stacked sleep inverter, as shown in Figure 18, operate similar to the sleep transistors in the sleep transistor technique. These sleep transistors are controlled by using control signals called 'Sleep' and 'SleepBar'. These sleep transistors are ON during active mode and OFF during standby mode.

Table 4. Simulation results of 6T CNTFET SRAM cell

Sl. No.	Parameter	Data-retention Sleep Transistor Technique	Forced Stack Technique	Sleep Transistor Technique	Base Case
1	Leakage Power (W)	1.89e-09	3.73e-09	0.325e-09	6.85e-09
2	Dynamic Power (W)	4.639e-07	4.631e-07	4.635e-07	4.63e-07
3	Read Delay (s)	7.95e-12	11.92e-12	7.81e-12	7.43e-12
4	Write Delay (s)	7.24e-12	10.3e-12	6.9e-12	6.5e-12
5	SNM (mV)	73.81	193	32.23	195
6	Area (nm^2)	17920	15360	16640	15360

Figure 18. Inverter with stacked sleep technique

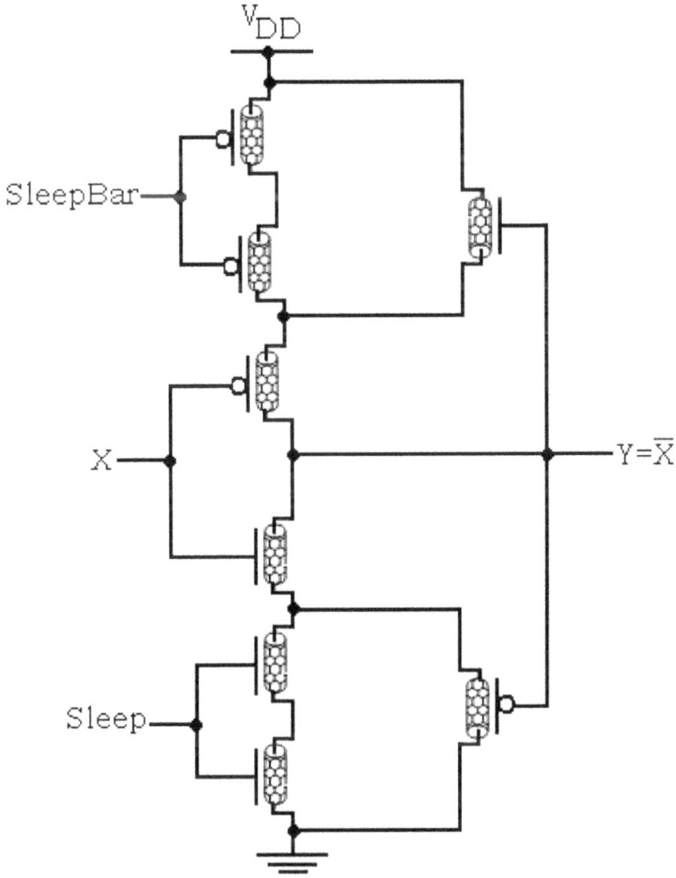

In active mode the control signals Sleep = '1' and SleepBar = '0' are asserted, and thus all sleep transistors are turned ON. Because of parallel transistors connected across stack of sleep transistors, delay is reduced in this proposed stacked sleep technique compared to forced stack technique.

In standby mode the control signals Sleep = '0' and SleepBar = '1' are asserted, and so that both of the sleep transistors are turned OFF. Although the sleep transistors are turned OFF during standby mode, this stacked sleep structure maintains exact logic state; this is because of parallel connected transistors. This technique reduces leakage power in two ways. In one way, leakage power is reduced due to the use of high-V_{TH} to sleep transistors. In second way, leakage power is reduced due to the connection of two OFF transistors in stack, i.e., due to stack effect. Because of these two effects, this stacked sleep structure achieves ultra-low leakage power consumption during standby mode while retaining exact logic state. The overhead of this technique is increase of area.

1.3.5.3 *Stacked Sleep 6T CNTFET SRAM Cell*

This technique is applied to 6T CNTFET SRAM cell as shown in Figure 19. This technique uses two stacked sleep transistor in V_{DD} and two stacked sleep transistor in ground. Leakage reduction in this technique occurs in two ways. First, the stack effect of sleep transistors and second the sleep transistor effect.

It is well known that PCNTFET transistors are not efficient at passing GND; similarly, it is well known that NCNTFET transistors are not efficient at passing V_{DD}. This stacked sleep technique uses PCNTFET transistor in GND and NCNTFET transistor in V_{DD} for maintaining the exact logic state during standby mode.

The stacked sleep transistors M7, M8 and M10, M11 are ON in active mode and OFF during standby mode. The sizing of a CNTFET is equivalent to adjusting the number of tubes. Since the mobility of NCNTFET and the mobility of PCNTFET carriers inside CNTs are identical, the minimum size is 1 for both PCNFET and NCNFET. This stacked sleep 6T CNTFET SRAM cell uses 1 tube for PCNTFET driver transistors (M2, M4) 2 tubes for NCNTFET access transistors (M5, M6) and 3 tubes for NCNTFET driver transistors (M1, M3). One tube is used for both the NCNTFET (M10 and M11) and PCNTFET (M7 and M8) stacked sleep transistors. The extra two parallel transistors used for maintaining the logic state during standby mode also uses one tube for both NCNTFET (M9) and PCNTFET (M12) transistors.

1.3.5.3.1 *Operation of Stacked Sleep 6T CNTFET SRAM Cell*

During standby mode Sleep='0' and SleepBar='1' are asserted. Assume that the stacked sleep 6T CNTFET SRAM cell stores logic '1' value i.e., Q='1'. This stacked sleep technique uses this Q value of '1' and an NCNTFET (M9) to maintain Q value equal to '1' when in standby mode. As shown in the Figure 19 an additional NCNTFET M9 is placed in parallel to the PCNTFET stacked sleep transistors (M7 and M8) connects the pull up network to V_{DD}. When in standby mode this NCNTFET (M9) is the only source of V_{DD} to the pull up network since the sleep transistors (M7 and M8) are turned OFF. Similarly to maintain a Q value of '0' in standby mode, this stacked sleep technique uses the Q value of '0' and a PCNTFET (M12) to maintain Q value equal to '0'.

Figure 19 shows an additional PCNTFET (M12) is placed in parallel to the pull down NCNTFET sleep transistors M10 and M11. When in standby mode, this PCNTFET (M12) is the only source of GND to the pull down network since the stacked sleep NCNTFETs M10 and M11 are turned OFF. Another advantage of this stacked sleep technique is that it uses high-V_{TH} for CNTFETs M7-M12 for more

Figure 19. Stacked sleep 6T CNTFET SRAM cell

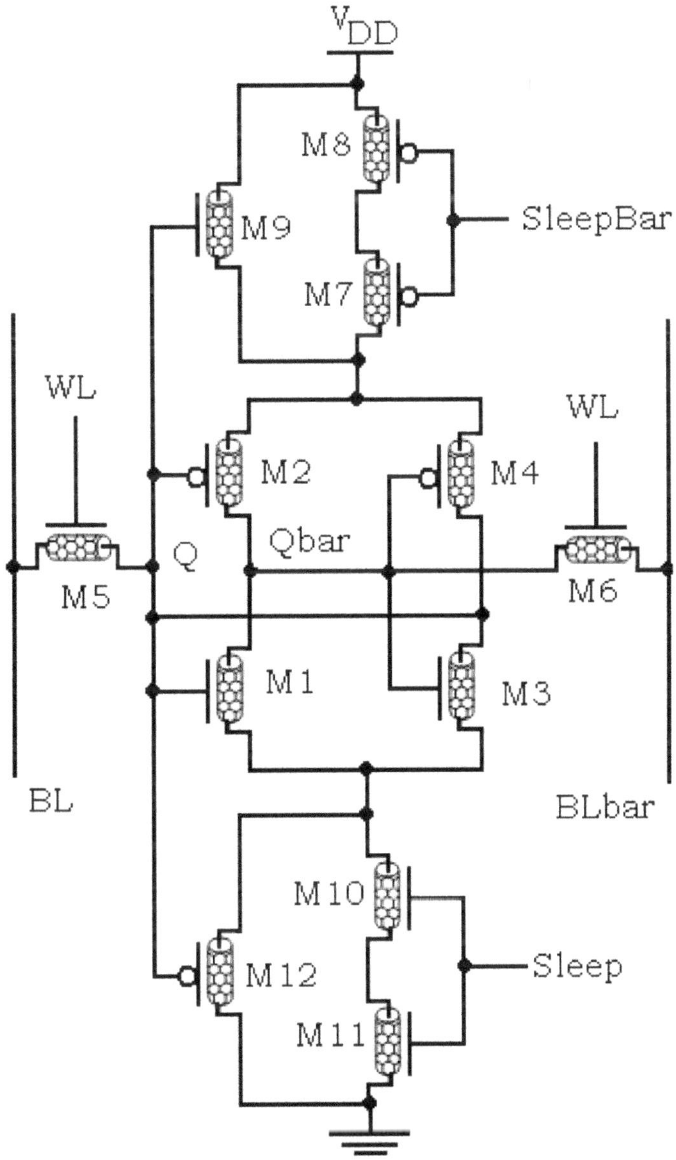

effective leakage reduction. Another advantage of parallel connected CNTFETs M9 and M12 in addition to retaining the state is that they reduces delay in active mode because of parallel connection between stacked sleep CNTFETs (M7, M8 and M10, M11) and these reduces resistance.

1.3.5.4 *Simulation Results and Discussions*

The 6T CNTFET SRAM cell is designed by applying this stacked sleep technique. This circuit is simulated using HSPICE by using Stanford CNTFET model at 32nm feature size (http://nano.stanford.edu/model.php?id=23). The simulated results of this stacked sleep 6T CNTFET SRAM cell are tabulated in Table 5. These results are compared to other and existing leakage reduction techniques in terms of leakage power, dynamic power, read delay, write delay, SNM and area.

This technique reduces leakage power by a large amount of 94.61% compared to base case 6T CNTFET SRAM cell, in addition to saving the state. There is, however, a 25% increase in cell area as compared to base case 6T CNTFET SRAM cell. This stacked sleep 6T CNTFET SRAM cell is better compared to data-retention sleep transistor 6T CNTFET SRAM cell due to its better leakage reduction. It is better compared to forced stack 6T CNTFET SRAM cell due to its better leakage reduction and also better than sleep transistor 6T CNTFET SRAM cell due to its state saving.

Compared to all these four 6T CNTFET SRAM cells and base case 6T CNTFET SRAM cell, this stacked sleep 6T CNTFET SRAM cell is the best option, because it retains the state as well as reduces the leakage power by a large value of 94.61% compared to base case 6T CNTFET SRAM cell and 80.47% compared to data-retention 6T CNTFET SRAM cell, 90.11% compared to forced stack CNTFET SRAM cell but compared to sleep transistor CNTFET SRAM cell the leakage reduction is 13.53% less. The overhead in this stacked sleep 6T CNTFET SRAM cell compared to base case is the increase of area by 25%. If area is not a constraint this stacked sleep 6T CNTFET SRAM cell is best in terms of leakage power saving with state saving capability.

Table 5. Simulation results of 6T CNTFET SRAM cell with all proposed and existing techniques

Sl. No.	Parameter	With Stacked Sleep Technique	With data-Retention Sleep Transistor Technique	With Forced Stack Technique	With Sleep Transistor Technique	Base Case
1	Leakage Power (W)	0.369e-09	1.89e-09	3.733e-09	0.325e-09	6.85e-09
2	Dynamic Power (W)	4.64e-07	4.6390e-07	4.631e-07	4.635e-07	4.63e-07
3	Read Delay (s)	9.47e-12	7.95e-12	11.92e-12	7.81e-12	7.43e-12
4	Write Delay (s)	8.61e-12	7.24e-12	10.3e-12	6.9e-12	6.5e-12
5	SNM (mV)	157	73.81	193	32.23	195
6	Area (nm^2)	19200	17920	15360	16640	15360

1.3.6 Results, Conclusions and Future Scope

1.3.6.1 Results

Therefore, this chapter focused on the design of a low power and high-performance SRAM cells using CNTFETs. The achievements of this chapter are described below:

A 6T SRAM cell is designed using CNTFET technology, called 6T CNTFET SRAM cell, and compared it with equivalent CMOS cell in terms of leakage power, dynamic power, read delay, write delay and SNM. All these results are given in Table 1.

The basic principle in sleep transistor 6T CNTFET SRAM cell is that simply cut-off the power supply when it is not in use i.e., when it is in standby mode. This sleep transistor 6T CNTFET SRAM cell is compared to base case 6T CNTFET SRAM cell in terms of leakage power, dynamic power, read delay, write delay, SNM and area. These simulated results are tabulated in Table 1.

The basic principle in forced stack 6T CNTFET SRAM cell is that exploiting stack effect, i.e., forcing a single OFF transistor into a stack of two-OFF transistors. Leakage current through series-connected stack of transistors, with more than one device OFF, is at least an order of magnitude smaller than that through single OFF device. This forced stack 6T CNTFET SRAM cell is compared to sleep transistor 6T CNTFET SRAM cell and base case in terms of leakage power, dynamic power, read delay, write delay, SNM and area and these simulated results are tabulated in Table 1.

The drawback of sleep transistor 6T CNTFET SRAM cell, is that during standby mode, the virtual ground node may get charged to V_{DD} or the virtual supply node may get discharged to '0', because of this the stored bit in this cell may get destroyed, so this sleep transistor 6T CNTFET SRAM cell cannot retain the state during standby mode. This data-retention sleep transistor avoided this drawback of sleep transistor technique by simply connecting virtual supply or virtual ground to fixed data retention voltage. This technique is applied to 6T CNTFET SRAM cell, named data-retention 6T CNTFET SRAM cell. This data-retention 6T CNTFET SRAM cell is compared to forced stack 6T CNTFET SRAM cell, sleep transistor 6T CNTFET SRAM cell and base case in terms of leakage power, dynamic power, read delay, write delay, SNM and area and these simulated results are tabulated in Table 4.

In this technique sleep transistors are stacked, so that the advantages both sleep and stack effect are utilized. Also this technique avoided drawbacks of sleep transistor and forced stack technique. This technique retained the stored cell state during standby mode and also largely reduced leakage power by using high-V_{TH} for sleep transistors. This stacked sleep technique is applied to 6T CNTFET SRAM cell, named stacked sleep 6T CNTFET SRAM cell. This stacked sleep 6T CNTFET

SRAM cell is compared to another data-retention 6T CNTFET SRAM cell, forced stack 6T CNTFET SRAM cell, sleep transistor 6T CNTFET SRAM cell and base case in terms of leakage power, dynamic power, read delay, write delay, SNM and area and these simulated results are tabulated in Table 5.

All these 6T CNTFET SRAM cells are designed and simulated by using **Synopsys HSPICE** circuit simulator with Stanford CNTFET model at 32nm technology at power supply V_{DD} of 0.9V.

CONCLUSION

As a summary, this chapter provided low power solutions for VLSI system designers. Especially, it focused on leakage power reduction solutions. In this chapter circuit-level leakage reduction techniques, named sleep transistor, forced stack, data-retention sleep transistor and stacked sleep are applied to 6T CNTFET SRAM cell to assess their performance. All these techniques performance is assessed in terms of SRAM metrics such as leakage power, dynamic power, read delay, write delay, SNM and area.

The following points are the conclusions arrived at in this chapter:

- The optimum chirality vector is found to be **(19, 0)** to achieve the best performance in terms of power consumption and stability of a 6T CNTFET SRAM cell. The leakage power in 6T CNTFET SRAM cell is reduced by **93.33%** and Static Noise Margin (SNM) indicates stability increased by **57.25%** compared to CMOS 6T SRAM cell.
- The sleep transistor 6T CNTFET SRAM cell reduces the leakage power by **95.25%** compared to the base case 6T CNTFET SRAM cell but at the cost of **6.15%** write delay and **8.33%** area. This cell is the best in terms of leakage power reduction but the main limitation of this cell is that it cannot retain its state during standby mode, as it is important in memories. This is because of its lowest SNM value of **32.22 mV** compared to base case SNM of **195 mV**.
- The forced stack 6T CNTFET SRAM cell reduces the leakage power by **45.5%** compared to base case 6T CNTFET SRAM cell in addition to state saving. Also observed that compared to base case, the forced stack 6T CNTFET SRAM cell occupies same area but with a dramatic increase of delay by **60.43%**. The leakage power reduction of this is small compared to other cells.
- The data-retention sleep transistor 6T CNTFET SRAM cell in addition to retaining the state, it also reduces leakage power by **72.4%** compared to base case 6T CNTFET SRAM cell and **49.32%** compared to other state saving forced stack 6T CNTFET SRAM cell. The area overhead in this method

compared to base case 6T CNTFET SRAM cell is **16%**. This data-retention sleep transistor 6T CNTFET SRAM cell is better compared to forced stack 6T CNTFET SRAM cell due to its more leakage power reduction and also better than sleep transistor 6T CNTFET SRAM cell due to its state saving.

- Another design is the stacked sleep 6T CNTFET SRAM cell. This design largely reduces leakage power by **94.61%** compared to base case 6T CNTFET SRAM cell, in addition to saving the state. There is, however, a **25%** increase in cell area as compared to base case 6T CNTFET SRAM cell. This stacked sleep 6T CNTFET SRAM cell is better compared to data-retention sleep transistor 6T CNTFET SRAM cell due to its better leakage reduction. It is better compared to forced stack 6T CNTFET SRAM cell because of its better leakage reduction. It is also better than sleep transistor 6T CNTFET SRAM cell due to its state saving.

The first data-retention sleep transistor 6T CNTFET SRAM cell, retains the state and reduces the leakage power by **72.4%** compared to base case 6T CNTFET SRAM cell and **49.32%** compared to other state saving forced stack 6T CNTFET SRAM cell. The area overhead in this method compared to base case 6T CNTFET SRAM cell is **16%**.

This chapter concluded that, compared to all these four 6T CNTFET SRAM cells and base case 6T CNTFET SRAM cell, the stacked sleep 6T CNTFET SRAM cell is the best option, because it retains the state as well as reduces the leakage power by a large value of **94.61%** compared to base case 6T CNTFET SRAM cell, **80.47%** compared to data-retention 6T CNTFET SRAM cell, **90.11%** compared to forced stack CNTFET SRAM cell. The leakage power of stacked sleep 6T CNTFET SRAM cell is very small and almost same as sleep transistor 6T CNTFET SRAM cell. The overhead in this stacked sleep 6T CNTFET SRAM cell compared to base case is the increase of area by **25%**. If area is not a constraint this stacked sleep 6T CNTFET SRAM cell is best in terms of leakage power saving as well as the state saving capability.

These designs can be used in low-power memory designs for any digital system.

SCOPE FOR FUTURE WORK

This section provides some directions for future research work based on the material presented in this chapter.

This work should be extended to higher CNTFET technologies such as 22nm, 16nm or beyond.

FinFET-based circuits are attractive for low-power, low voltage applications. FinFET-based SRAM (Cakici, Kim & Roy, 2007) has shown to be very effective in reducing the short channel effects. One interesting extension of our research of ultra-low leakage CNTFET SRAM cells design, is to evaluate the efficacy of the SRAM cell with these proposed methods based on the FinFETs.

Due to its desirable electronic and mechanical properties, many numbers of scientists are researching the potential of graphene in ICs. A simple graphene IC has been fabricated by IBM. Researchers have also demonstrated graphene analogue transistors operating at 155 GHz. People at the microelectronics research center in Texas, are working on creating digital circuits made from graphene. Research is going on to study the fundamental advantages of using graphene in electronics, and the potential for creating the first commercially available graphene digital microprocessors by 2023 (C. Xu, H. Li, & K. Banerjee, 2009). So the future research can focus on design of SRAM memory using GNRFETs.

REFERENCES

Akturk, G., Pennington, G., Goldsman, N., & Wickenden, A. (2007). Electron transport and velocity oscillations in a carbon nanotube. *IEEE Transactions on Nanotechnology*, *6*(4), 469–474. doi:10.1109/TNANO.2007.896968

Bachtold, P. (2001). Logic Circuits with Carbon Nanotube Transistors. *Science*, *294*(5545), 1317–1320. doi:10.1126cience.1065824 PMID:11588220

Berkeley Predictive Technology Model (BPTM). (n.d.). Available http://wwwdevice.eecs.berkeley.edu/˜ptm/

Cakici, T., Kim, K., & Roy, K. (2007). FinFET Based SRAM Design for Low Standby Power Applications. *8th International Symposium on Quality Electronic Design*, 127-132. 10.1109/ISQED.2007.76

Chandrakasan, P., Sheng, S., & Brodersen, R. W. (1992). Low-Power CMOS Digital Design. *IEEE Journal of Solid-State Circuits*, *27*(4), 473–484. doi:10.1109/4.126534

Chen, D., Chilstedt, S., Dong, C., & Pop, E. (2010). *What Everyone Needs to Know about Carbon-Based Nano circuits*. DAC.

de Alba Rosano, M. R., & Garcia-Garcia, A. D. (2006). Measuring leakage power in nanometer CMOS 6T SRAM cells. *Proc. IEEE Int. Conf. Reconfigurable, Comput. FPGA's*.

Deng, J., & Wong, H.-S. P. (2006). A Circuit-Compatible SPICE model for Enhancement Mode Carbon Nanotube Field Effect Transistors. *Proc. Intl. Conf. Simulation of Semiconductor Processes and Devices*, 166 – 169. 10.1109/SISPAD.2006.282864

Deng, J., & Wong, H.-S. P. (2007). A compact spice model for carbon-nanotube field-effect transistors including nonidealities and its application. *IEEE Transactions on Electron Devices, 54*(12), 3186–3194. doi:10.1109/TED.2007.909030

Franklin, A. D., Luisier, M., Han, S.-J., Tulevski, G., Breslin, C. M., Gignac, L., ... Haensch, W. (2012). Sub-10 nm Carbon Nanotube Transistor. *Nano Letters, 12*(2), 758–762. doi:10.1021/nl203701g PMID:22260387

Hashempour, H., & Lombardi, F. (2008). Device model for ballistic CNFETs using the first conducting band. *IEEE Design & Test of Computers, 25*(2), 178–186. doi:10.1109/MDT.2008.34

ITRS. (2009). *Process Integration, Devices, and Structures*. Retrieved from http://www.itrs.net/Links/2009ITRS/2009Chapters_2009Tables/2009_PIDS.pdf

Javey, Wang, Kim, & Dai. (2003). Advancements in complementary carbon nanotube field-effect transistor. *Proc. 2003 IEEE Int. Electron Devices Meeting*, 31.2.1–31.2.4. 10.1109/IEDM.2003.1269387

Johnson, M., Somasekhar, D., Chiou, L.-Y., & Roy, K. (2002). Leakage control with efficient use of transistor stacks in single threshold CMOS. *IEEE Trans. VLSI Systems., 10*(1), 1–5. doi:10.1109/92.988724

Kim, Nan, & Choi. (2010). Power Gating for Ultra-Low Voltage Nanometer ICs. *IEEE ISCAS*.

Kuresh & Hasan. (2009). Performance comparison of CNFET- and CMOS-based 6T SRAM cell in deep submicron. *Microelectronics Journal, 40*(6), 979-982.

Lin, Kim, & Lombardi. (2008). A New SRAM Cell Design Using CNTFETs. *International SoC Design Conference, 2008, 1*, I-168 - I-171.

Lin, S., Kim, Y.-B., & Lomabrdi, F. (2011). The CNTFET-Based Design of Ternary Logic Gates and Arithmetic Circuits. *IEEE Transactions on Nanotechnology, 10*(2), 217–225. doi:10.1109/TNANO.2009.2036845

Lin, Y., Appenzeller, J., Knoch, J., & Avouris, P. (2005). High-performance carbon nanotube field-effect transistor with tunable polarities. *IEEE Transactions on Nanotechnology*, *4*(5), 481–489. doi:10.1109/TNANO.2005.851427

Meindl, J. D. (1995). Low power Microelectronics: Retrospect and Prospect. *Proceedings of the IEEE, 83*(4), 619–635. doi:10.1109/5.371970

Moradinasab, M., & Fathipour, M. (2009). Stable, low power and high performance SRAM based on CNFET. *10th International Conference on Ultimate Integration of Silicon (ULIS 2009),* 317 – 320.

Pant, P., De, V., & Chstterjee, A. (1997). Device-Circuit Optimization for Minimal Energy and Power Consumption in CMOS Random Logic Network. *ACM/IEEE Design Automation Conf.*, 403-408.

Patil, N., Lin, A., Zhang, J., Wong, H.-S. P., & Mitra, S. (2009). Digital VLSI logic technology using carbon nanotube FETs: Frequently asked questions. *Proc. IEEE Design Autom. Conf.*, 304–309. 10.1145/1629911.1629995

Qe, H., Kim, Y.-B., & Choi, M. (2012). *A High Speed Low Power Modulo 2^n+1 Multiplier Design Using Carbon-nanotube Technology. IEEE International Midwest Symposium on Circuits and Systems (MWCAS),* Boise ID.

Rahman, Guo, Datta, & Lundstrom. (2003). Theory of ballistic nanotransistors. *IEEE Trans. Electron Devices, 50*(10), 1853–1864.

Stanford University CNFET Model website. (n.d.). Available: http://nano.stanford.edu/model.php?id=23

Wang, W., & Choi, K. (2010). Novel curve fitting design method for carbon nanotube SRAM cell optimization. *IEEE International Conference on Electro/Information Technology (EIT).*

Wang, W., Yu, Z., & Choi, K. (2011). *High SNM 6T CNFET SRAM Cell Design Considering Nanotube Diameter and Transistor Ratio*. IEEE EIT. doi:10.1109/EIT.2011.5978618

Wei, Roy, & De. (2000). Low Voltage Low Power CMOS Design Techniques for Deep Submicron ICs. *13th International Conference on VLSI Design*, 24 – 29.

Xu, C., Li, H., & Banerjee, K. (2009). Modeling, analysis and design of grapheme nano-ribbon interconnects. *IEEE Transactions on Electron Devices, 56*(8), 1567–1578. doi:10.1109/TED.2009.2024254

Yu, Z., Chen, Y., Nan, H., Wang, W., & Choi, K. (2011). *Design of Novel Low Power 6T CNFET SRAM Cell Working in Sub-Threshold Region*. IEEE EIT.

Section 3
CNTs as Interconnects

Chapter 7
CNT as Interconnects

Karmjit Singh Sandha
Thapar Institute of Engineering and Technology, Patiala, India

ABSTRACT

The chapter will start with brief introduction to the interconnects and its importance in an integrated circuit at deep sub-micron technology nodes. The brief discussion about the concept of scaling, interconnects models, and material in use are presented. The limitations of conventional materials at scaled down technology nodes will be discussed next. The focus of the chapter is to present the electrical equivalent circuit model to estimate the impedance parameters of SWCNT bundle and MWCNT bundle as interconnects at different nano-scaled technology nodes for global level interconnect length. Using ESC model of SWCNT, MWCNT, and copper, the performance comparative analysis for delay and power delay product (PDP) will be presented for different interconnect lengths at nano-scaled technology nodes. Finally, the chapter summary and conclusion will be written at the end of the chapter.

1. INTRODUCTION

An integrated circuit is created by combining thousands of transistors or devices or modules onto a single chip. After the introduction of Very Large Scale Integration (VLSI) circuits in 1970s, numerous functions can be performed on single IC chips. The electronics industry has shown a phenomenal growth in recent years. The high computational speed and efficient processing power is the driving force for technology up gradations. The number of complex functions is increasing in small package, so there has been steady increase in the level of integration as well. The beauty of integrated circuit lies in the ultimate performance in all aspects like chip area, cost, speed and power of all interconnected modules. So, interconnect issues

DOI: 10.4018/978-1-7998-1393-4.ch007

are very important at scaled technology nodes. If the technology somehow fails to have a good interconnect, then total performance of IC will degrade due to high parasitic resistance and capacitances leading to even higher delay than transistors or logic gates. (Bakoglu & Meindl, 1985; Steinhögl et al., 2005)

2. VLSI INTERCONNECTS

Interconnects are thin conducting paths used to establish the electrical links between two or more than two nodes of the electrical circuit within the Integrated Circuits (IC). Different conducting materials have been used as interconnects in ICs. In early days the aluminum was used as interconnect material but due to its large resistivity at micro-scaled technology nodes, the copper was used as alternative material for interconnects in IC design. As technology is scaled down to nanometer technologies, the resistance of copper interconnects increased rapidly due to its small mean free path (MFP) (Kaushik & Majumder, 2015; Gholipour & Masoumi, 2012; Naeemi & Meindl, 2009; Srivastava et al., 2009; Burke, 2002). MFP of electrons is due to the combined effects of scattering (grain boundary and surface) and electromigration. Due to these disadvantages of copper as interconnect material at advanced technologies, the conventional copper material is replaced by other new materials. As the technology nodes are scaled down, the density of the devices is increasing and long interconnect lengths are required to interface all the devices. Therefore, the performance of an IC is mainly on basis of current carrying capacity and parasitic such as resistance, inductance and capacitance of the interconnect material (Das, Majumder, & Kaushik, 2014; Singh & Raj, 2015; Sandha & Sharma, 2018; Hosseini & Shabro, 2010; Pop et al., 2007). As technology nodes are scaled down, the device dimensions and supply voltages of ICs are also scaled down, the interconnect dimensions are also required to be scaled down to synchronized the dimensions (Singh & Raj, 2015). With advancement of technology nodes, more functions are to be incorporated in VLSI chips. Therefore, the required interconnects length is exponentially increasing to connect millions of active devices in with in an IC. The basic structure of Driver Interconnects Load (D-I-L) is shown in Figure 1.

In DIL, a CMOS based inverter is used to drive interconnect with a capacitive load as shown in Figure 1. To evaluate the performance of interconnect, the equivalent parameters of used material will be used as R-L-C circuits model. These impedance parameters of interconnects are depending on the technology dependent cross-sectional dimensions and properties of the material used as interconnects. (Steinhögl et al., 2005; Majumder, Kaushik & Manhas, 2014)

Figure 1. Basic DIL structure of VLSI interconnects

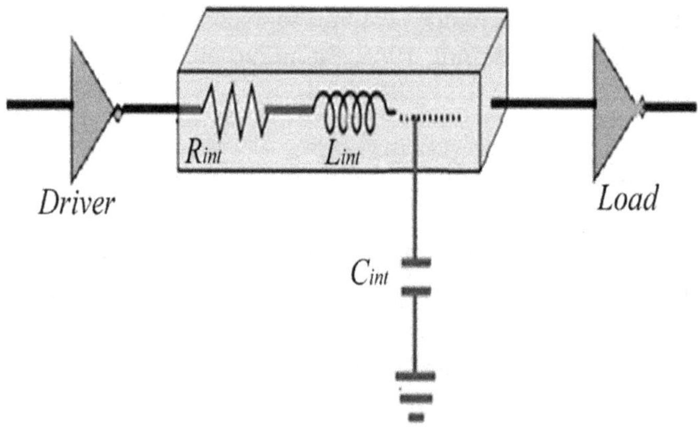

On the basis of cross-sectional dimensions and interconnect length, interconnects in ICs can be classified into following three types:

- Local Interconnect.
- Intermediate interconnects.
- Global interconnects.

Local Interconnects: Local interconnects having small interconnects length ranging from few μm to tens of μm and used to interface immediate nodes. These are usually interface the terminals of CMOS device in CMOS technology i.e. gates, sources and drains of CMOS devices. The local interconnects can also be used as vias which are used to connect gate electrode material for CMOS technology. Since, the length of local interconnect is small (few μm), so, they can afford the high resistivity materials as interconnects as compare to long interconnects. (Das, Majumder, & Kaushik, 2014; Singh, & Raj, 2015)

Intermediate Interconnects: Intermediate interconnects deals with both short and long interconnects length. This type of interconnects are also known as semi-global interconnects. The different devices within a block are interfaced through intermediate interconnects. Intermediate interconnects are also used to interface two or more blocks with typical interconnect lengths up to 500um. (Das, Majumder, & Kaushik, 2014; Singh, & Raj, 2015)

Global Interconnects: The long interconnects length (> 500μm) used for power supply, common ground and clock are known as global interconnects. In long interconnects, the power, delay and power delay product (PDP) are the key factors to evaluate the performance of interconnect. Therefore, global level interconnects should have to use low resistive metallic materials as they often travel over large distances to provide connection between different parts of the circuit and different devices. (Rossnagel & Kuan, 2004)

The cross-sectional view shows that with technology scaling down, local interconnects are used to connect the nearby nodes at the lowest level of the hierarchy followed by intermediate level interconnects used to connect the devices within the blocks, Global interconnects are generally used for long interconnects lengths to connects various functional blocks, power supply, clocks and grounds.

2.1. Importance of Scaling

Scaling process in VLSI circuits design, related to systematic reduction of device dimensions as permitted by technology while geometric ratios remains preserved. The designing of high performance and high density VLSI chip design, needs the high packing density of CMOS based devices. Therefore, the size of these devices needs to be kept small to meet the required area considerations. The channel length and width of CMOS transistor are scaled down as the technology nodes scaled down. As the device dimensions are scaled down, the dimensions of interconnects are also needed to be scaled down (Steinhögl et al., 2005). The cross sectional view of predicted interconnects length of present and future scenario is shown in Fig. 15.2. It is shown in Fig. 15.2 that interconnects can be scaled down with different scaling factor at scaled down technology nodes. Similarly, wire lengths are also needs to be scaled for both local and global level interconnects length, for future IC design.

To estimate the performance of an integrated circuits, the effect of scaling on interconnects are required to be considered. The effects of scaling on local interconnects is negligible small as most of the performance estimation parameters are device dependent. Due to large impedance parasitic of long interconnects, the performance of interconnects dominates over the total performance of an IC. For long interconnects, if the device dimensions are scaled with a scaling factor 1/S then, the propagation delay will be increased by S2* S_C^2, where S_C^2 is scaling factor for size of chip. (Bakoglu, & Meindl, 1985; Steinhögl, et al., 2005; Li et al., 2009; Im et al., 2005)

2.2. Significance of Repeaters for Long Interconnects

Interconnects can be categories into three types on the basis of their lengths as describes in previous sections. The local interconnects are used as small interconnects wires, the impedance parasitic of interconnects are small compared to the impedance parasitic of driver transistor. The impedance parameters of interconnects increases linearly

Figure 2. Cross-sectional view of present and future interconnect length required

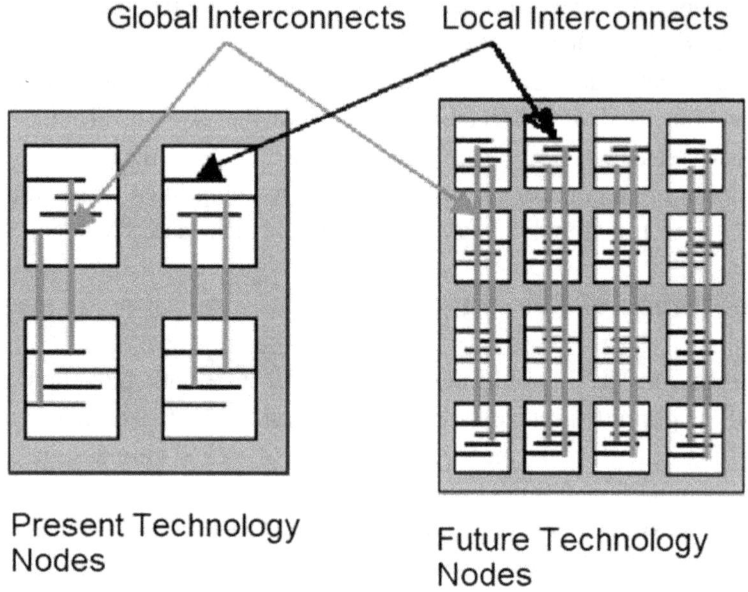

as interconnects length increases. Thus, the propagation delay of interconnects also increases with increase in length. The impedance parameters of the long interconnect becomes larger as compare to driver transistor. Therefore, these long interconnects can be divided into smaller subsections and these subsections can be separated with CMOS invertor called repeaters to decrease the overall delay of long interconnects. The propagation delay (RC time constant) used to estimate the performance of VLSI circuits and it increases sharply as interconnect dimensions are scaled down with advanced technology nodes and affect the speed of VLSI circuits. Therefore, to improve the performance in terms of propagation delay and speed of the circuit, optimum number repeaters and optimum sized repeaters can be used. (Bakoglu, & Meindl, 1985; Li et al., 2008; Li et al., 2009; Litoria, Sandha, & Kansal, 2017)

3. DIFFERENT MATERIALS AS INTERCONNECTS

The section presents the properties, advantages and disadvantages of different material used as VLSI interconnects material at different technology nodes

3.1. Aluminum as Interconnects

In earlier technology nodes, the aluminum was used as interconnect material due to its excellent characteristics such as good conductivity, good ohmic contact with silicon and adheres on silicon dioxide. As technology scaled down to micro-scaled level, interconnects current density increases and causes electromigration. Electromigration referred as the transfer of energy between the conducting electrons and the diffusing metal atoms, transport of material thus takes place by the gradual movement of ions. As the technology node further scaled down, the effect of electromigration increases. Moreover, the aluminum has the resistivity in the order of 2.7 μΩ–cm which is very higher than the resistivity of copper (1.7 μΩ–cm). Therefore, at micro-scaled technology nodes, there is need of high current density but due to its large resistivity and electromigration, high current density cannot be achieved. Due to these disadvantages, copper was used as an alternative interconnects material at micro-scaled technology nodes. (Rossnagel, & Kuan, 2004).

3.2. Copper as Interconnects

Aluminum was replaced with copper as an alternative material for interconnects, due to the shortcomings of aluminum at micro-scaled technology nodes. The copper has high current density, low resistivity and more immune to electromigration compared to aluminum. Copper also has higher electrical conductivity and melting point (1357K) compared to aluminum (933K) making it thermally more stable. (Singh & Raj, 2015). Comparative analysis of some properties of copper and other conducting materials is shown Table 1.

Some other properties of copper as shown in Table 15.1 which make more suitable alternative compared to aluminum at micro-scaled technology nodes for VLSI interconnects. However, as technology is scaled down to nanometer i.e. less than 45nm, the resistance of copper interconnects increased rapidly because of its

Table 1. Comparison of different properties of conducting materials

Properties / Materials	Resistivity	Thermal Conductivity	Melting Point	Corrosion in Air	Adhesion to SiO2
Copper (Cu)	1.67	3.98	1085	Poor	Not good
Silver (Ag)	1.59	4.25	962	Poor	Not good
Gold (Au)	2.35	3.15	1064	Excellent	Not good
Aluminum (Al)	2.66	2.38	660	Good	Good
Tungsten (W)	5.65	1.74	3387	Good	Not good

small MFP. The MFP is the average distance travelled by an electron before its collision. The average MFP of copper is approximately equal to 45nm. (Steinhögl et al., 2005; Li et al. 2009). The MFP of electrons is depending upon the grain boundary scattering, surface scattering and electromigration. Grain boundary is the interface between two crystallites in a polycrystalline material. Due to grain boundaries introduce electron scattering which leads to reduce the mobility of electron and hence, it tends to increase the thermal and electrical resistivity. As technology scaled down to nanometers (<45nm), the MFP of electron of copper becomes comparable to the cross-section dimensions of interconnects. Hence, there is more reflection of an electron at the surfaces cause the scattering. The scattered electrons emit or absorb phonon, introduce impurities in the copper. So, the total collisions rate with the surfaces scattering will suddenly increase and hence increase the resistance of copper at nano-scaled technology nodes, impose the serious challenges for interconnects delay and systems reliability.

The sudden rise in resistance has a considerable effect on the overall performance and reliability particularly for long interconnects of VLSI circuits design. In the years, CNT has been considered as an alternative material to copper for global level interconnects (Steinhögl et al., 2005; Li et al. 2009).

3.3. Carbon Nanotubes as Interconnects

CNTs have developed as future material for next generation VLSI technology nodes. CNTs are made by rolling up a carbon grapheme sheet into a cylindrical form. These cylindrical shaped carbon molecules have attractive thermal and electrical properties which make it more suitable in the field of electronics, nanotechnology, optics, material science and other fields of technology. In particular, CNTs have applications in field of VLSI interconnects due their extraordinary properties. CNTs have longer MFP (1000nm) and high current density ($>10^{10}$) as compared to copper (Rutherglen & Burke, 2009; Singh & Raj 2015; Sandha & Sharma 2018). A comparison of different properties of CNTs and copper is shown in Table 15.2.

3.4. Classification Of CNT

Carbon nanotubes are classified into different types according to their conductivity, chirality and structure. CNTs can be classified into two types on the basis of chiral indices (n,m): armchair and zigzag. When the any one chiral indices equals to zero, resultant structure is known as zigzag and for armchair chiral indices should be same i.e. m=n. Depending upon the direction of CNT in which they are rolled up, decide the conductivity, either metallic or semiconductor. The armchair structures are always metallic and zigzag structures are either metallic or semiconducting depends upon

Table 2. Comparison of different properties of Copper, SWCNT, MWCNT and Graphene

Properties\Material	Cu	SWCNT	MWCNT	Graphene
Maximum current density(A/cm^2)	10^6	10^{10}	10^{10}	10^9
Thermal Conductivity(10^3 .W/mK)	0.385	1.75-5.8	3	3-5
Mean free Path(nm)	40	$>10^3$	2.5X10^4	1X10^3
Melting Point (K)	1357	3800(graphite)		
Temperature coefficient of resistance (10^{-3}/K)	4	<1.1	-1.37	-1.47
Mechanical Strength	Poor	Fair	Fair	Fair
Process of Fabrication	Easy	Difficult	Difficult	Difficult

the chiral indices. For VLSI interconnects only metallic CNTs structure can be used because of their high electrical and thermal properties. Further, on the basis of its structures, CNTs are classified into two types:

- Single-Walled Carbon Nanotubes (SWCNT).
- Multi-Walled Carbon Nanotubes (MWCNT).

SWCNT is a graphene sheet rolled up to form a cylindrical shaped tube with identical diameter as shown in Fig. 15.3 (a). In interconnects, more than one parallel SWCNTs have to act as CNT bundle in which all the parallel SWCNTs are of identical diameter and number of conducting channels. Therefore, the impedance parameters of a bundle are calculated by considering them in a parallel combination.

In case of MWCNT bundle, more than on CNTs are concentrically inserted to each other as shown in Figure 3 (b). The diameters of different CNTs of a bundle

Figure 3. (a) Structure of SWCNT and (b) MWCNT

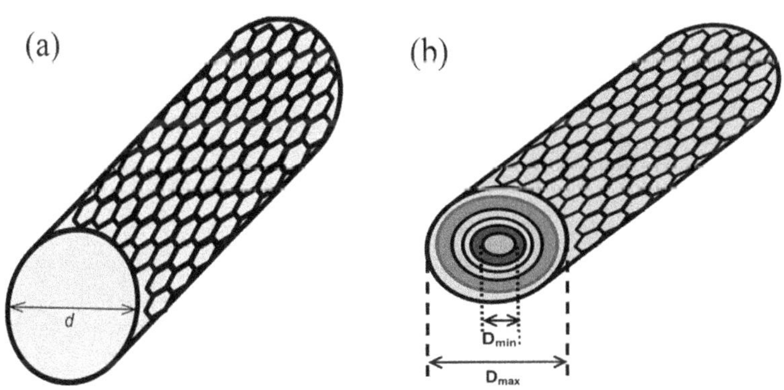

are larger than the diameter of SWCNT and one MWCNT bundle consists of large number of shells and conducting channels as shown in Fig. 15.3 (b). SWCNT and MWCNT have same current carrying capacity but later are easy to fabricate as compare to former because of its better control on the growing process. (Singh & Raj 2015; Kaushik & Majumdar 2015; McEuen, Fuhrer & Park 2012)

4. EQUIVALENT CIRCUIT MODEL FOR SWCNT

A SWCNT bundle consists of many cylindrical shaped carbon nanotubes with identical diameter. An individual shell of SWCNT bundle with small diameter (in the range of 1nm to 10nm) has equivalent electrical circuit model which consists of diameter, length and impedance parameters. Figure 4 (a) shows a schematic structure of SWCNT on ground plane with diameter d and separated the ground plane with distance y1. The equivalent impedance circuit model for a SWCNT is shown in Figure 4 (b).

4.1. Impedance Model of an SWCNT

The equivalent resistance of SWCNT bundle can be divided into three parts: quantum resistance (R_Q) (6.45kΩ), the electron-phonon scattering based resistance (R_S) (for length > MFP of CNT) and contact resistance (Rc).

The contact resistance (Rc) is due to contact of metal with CNT and it is dependent upon the fabrication process of CNT.

Figure 4. (a) Schematic structure of a SWCNT on ground plane (b) Equivalent impedance model of SWCNT

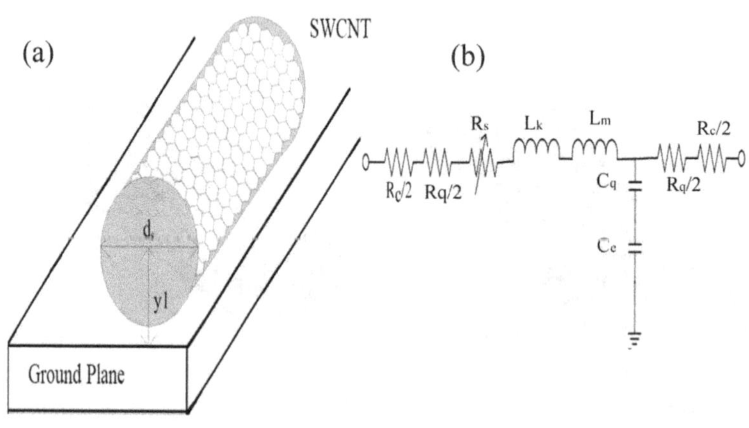

Therefore, the total resistance for each SWCNT tube is given by

$$R_{shell} = R_c + R_q + R_s.L$$
$$R_{ohcll} = 2k\Omega + \frac{h}{2e^2N} + \frac{h}{2e^2N}.\frac{L}{\lambda} \tag{15.1}$$

where $h/2e^2 = 12.9k\Omega$, N, L and λ are the number of conducting channels in a shell, interconnect length and mean free path (MFP) of every SWCNT respectively. It is revealed from Eq. (15.1) that MFP has an important role to play determine the total resistance of SWCNT. (Srivastava et al., 2009; Singh & Raj 2015; Majumder, Kaushik & Manhas 2014)

SWCNT is also having two types of inductances i.e. magnetic and kinetic inductance per unit length. The magnetic inductance is due to the magnetic field of current carrying isolated conductor with at a distance from ground as shown in Figure 4 (a). The magnetic inductance for each tube of SWCNT is as

$$L_M = \left(\frac{\mu}{2\pi}\right)\ln\left(\frac{y}{d}\right) \tag{15.2}$$

where μ is mobility of electron and y is the separation between isolated SWCNT to ground.

The kinetic inductance is due to kinetic energy stored in the every conducting channels of SWCNT when shell conducts and given as

$$L_K = \frac{h}{4N.v_f e^2} \tag{15.3}$$

where v_f is Fermi velocity of electron and e is electron charge and h is Planck constant.

The capacitance in SWCNT is also of two types: quantum capacitance (C_Q) and electrostatic capacitance (C_E). The quantum capacitance (C_Q) is due to quantum electrostatic charge stored in CNT during its conduction. (Singh & Raj, 2015) The quantum capacitance for an isolated SWCNT is given by

$$C_Q = \frac{4Ne^2}{hv_f} \tag{15.4}$$

The electrostatic capacitance (C_Q) of an isolated SWCNT is calculated by considering a SWCNT as a conducing wire having diameter d and placed above ground at a distance y as shown in Figure 4 (a) and given as

$$C_E = \frac{2\pi\varepsilon}{\ln\left(\frac{y}{d}\right)} \quad (15.5)$$

4.2. Impedance Model of SWCNT Bundle

When more than two such SWCNTs are placed in parallel to use it as VLSI interconnects is known as SWCNT bundle interconnect as shown in Figure 5 (a). The equivalent impedance circuit of SWCNT bundle is shown in Figure 5 (b).

The resistance of SWCNT bundle (R_{Bundle}) with similar length L and MFP (λ) can be considered as parallel and given by

$$R_{(Bundle)} = \frac{R_Q + R_S.L + R_{mc}}{n_{bundle}}$$
$$R_{(Bundle)} = \frac{\frac{h}{2e^2N} + \frac{h}{2e^2N}.\frac{L}{\lambda} + R_{mc}}{n_{Bundle}} \quad (15.5)$$

nBundle is total number of shells in a bundle and number of conducting channels in an individual shell can be given by

$$N \approx a.d + b \qquad if\ diameter > 3nm \quad (15.6)$$

where d is the diameter of an individual shell, a=6.12x 10^{-4} 1/(nmK) and b=1.275. If the diameter of an isolated SWCNT is less than 3nm than number of conducting channels can be calculated as (Singh & Raj 2015)

$$N \approx 2, \qquad if\ diameter < 3\ nm \quad (15.7)$$

The diameter of each shell of a SWCNT bundle is assumed to be identical and considered to be equal 1nm for the research work. Hence, the conducting channels

Figure 5. (a) Cross-sectional view of SWCNT bundle (b) Equivalent impedance model for SWCNT bundle

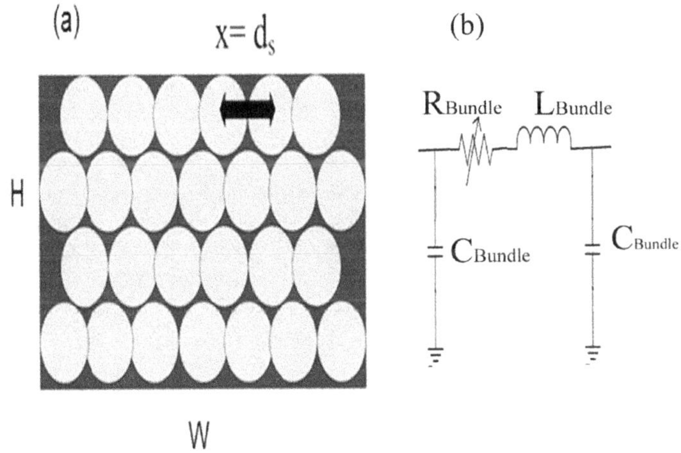

for each shell are considered equal to two. Therefore, the total resistance of SWCNT bundle can be re-written as

$$R_{(Bundle)} = \frac{\frac{h}{4e^2}\left(1+\frac{L}{\lambda eff(T)}\right)+R_c}{n_{Bundle}} \tag{15.8}$$

Where, nBundle in a bundle, depends upon number of rows ($n_{H(Bundle)}$) and number of columns ($n_{W(Bundle)}$) of the bundle. The parity of number of rows ($n_{H(Bundle)}$) of a bundle, decides the total number of SWCNTs in a bundle and total tubes for even number of rows in a bundle is given as

$$n_{Bundle} = n_{W(Bundle)} . n_{H(Bundle)} - \frac{1}{2} n_{H(Bundle)} \tag{15.9}$$

Similarly for odd number of rows, the total number of SWCNT tubes (n_{Bundle}) can be calculated as

$$n_{Bundle} = n_{W(Bundle)} . n_{H(Bundle)} - \frac{1}{2}(n_{H(Bundle)} - 1) \tag{15.10}$$

Where, $n_{W(Bundle)}$ and $n_{H(Bundle)}$ are number of columns and rows respectively for a SWCNT bundle are given as

$$n_{W(Bundle)} = \left(\frac{W-d}{x}\right) \tag{15.11}$$

$$n_{H(Bundle)} = \left(\left\lfloor \frac{H-d}{\left(\frac{\sqrt{3}}{2}\right)x} \right\rfloor + 1\right) \tag{15.12}$$

where x, H and W are space between adjacent tubes, height and width of bundle respectively as shown in Figure 5.(b).

Hence, if number of rows are even then the total number of SWCNT shells (nBundle) in a bundle given as

$$n_{Bundle} = \left\lceil \frac{W-d}{x} \right\rceil \left(\left\lfloor \frac{H-d}{\left(\frac{\sqrt{3}}{2}\right)x} \right\rfloor + 1\right) - \frac{1}{2}\left(\left\lfloor \frac{H-d}{\left(\frac{\sqrt{3}}{2}\right)x} \right\rfloor + 1\right) \tag{15.12}$$

Similarly, the SWCNT shells (n_{Bundle}) for odd number of rows is given as

$$n_{Bundle} = \left\lceil \frac{W-d}{x} \right\rceil \left(\left\lfloor \frac{H-d}{\left(\frac{\sqrt{3}}{2}\right)x} \right\rfloor + 1\right) - \frac{1}{2}\left(\left\lfloor \frac{H-d}{\left(\frac{\sqrt{3}}{2}\right)x} \right\rfloor\right) \tag{15.13}$$

The total inductance of bundle is also calculated by considering all SWCNTs in parallel combination and given by

$$L_{(Bundle)} = \left(\frac{L_M + L_K}{4n_{Bundle}} \right) \tag{15.14}$$

where LM and LK are magnetic and kinetic inductances of SWCNT and can be calculated by using Eqs.(15.2) and (15.3).

The total capacitance of SWCNT bundle is the parallel combination of bundle electrostatic capacitance $C_E^{(Bundle)}$ and bundle quantum capacitance C_Q^{Bundle} of SWCNT and can be given as

$$C_E^{(Bundle)} = 2\left(\frac{2\pi\varepsilon_{ox}\varepsilon_0}{\ln(S/d)} \right) + \left(\frac{W-d-2}{2} \right)\left(\frac{2\pi\varepsilon_{ox}\varepsilon_0}{\ln(\frac{S+W}{d})} \right) + 3\left(\frac{2\pi\varepsilon_{ox}\varepsilon_0}{\ln(S/d)} \right)\left(\frac{n_{H(bundle)}-2}{5} \right) \tag{15.15}$$

$$C_Q^{Bundle} = \left(\frac{2\times 2e^2}{hv_f} \right) n_{CNT} \tag{15.16}$$

$$C_{(Bundle)} = \left(\frac{C_E^{Bundle} . C_Q^{Bundle}}{C_E^{Bundle} + C_Q^{Bundle}} \right) \tag{15.17}$$

where v_f is Fermi velocity and S is distance between adjoining bundles. (Singh & Raj, 2015)

5. EQUIVALENT ELECTRICAL CIRCUIT MODEL FOR MWCNT

MWCNT bundle contains several shells like SWCNTs concentrically nested with different diameters one inside the other, as shown in Figure 6. The different shells of the MWCNT bundle have different diameters and vary from few nanometers to hundreds of nanometers depending on the technology nodes. The diameters of the outermost shell and innermost shell of MWCNT bundle are Douter and Dinner, respectively. The center of MWCNT bundle is separated from the ground plane with a distance Y. The spacing between two adjacent shells MWCNT bundle may vary from 0.3nm to 0.4nm for different configurations. The small adjacent distance leads towards to increase inter-shell capacitance which may effects the average power consumption of interconnects, whereas the large inter-shell distance increases the overall resistance of the bundle and effects the propagation delay. The work presents in the thesis has considered the distance of adjacent shells is equal to $\delta \approx 0.34$nm (van der Waals distance). (Singh & Raj, 2015)

5.1. Conducting Shells and Channels in a Bundle

The shells in a bundle are numbered from outermost shell to innermost shell as 1, 2,.,i...n. The number of shells in a bundle depends upon the outermost diameter and diameter ratio of the bundle, where diameter ratio is defined as ratio between the diameter of outermost shell to innermost shell of MWCNT bundle and given as D_{inner}/D_{outer}.

The outermost shell of MWCNT bundle is technology dependent parameter and diameter ratio (D_{inner}/D_{outer}) may vary from 0.35- 0.8 depending upon fabrication process. Therefore, total n number of shells in a bundle can be calculated as

$$n = 1 + \text{int}\left[\left(\frac{D_{outer} - D_{inner}}{2\delta}\right)\right] \tag{15.18}$$

Figure 6. Structure of MWCNT bundle on a ground plane

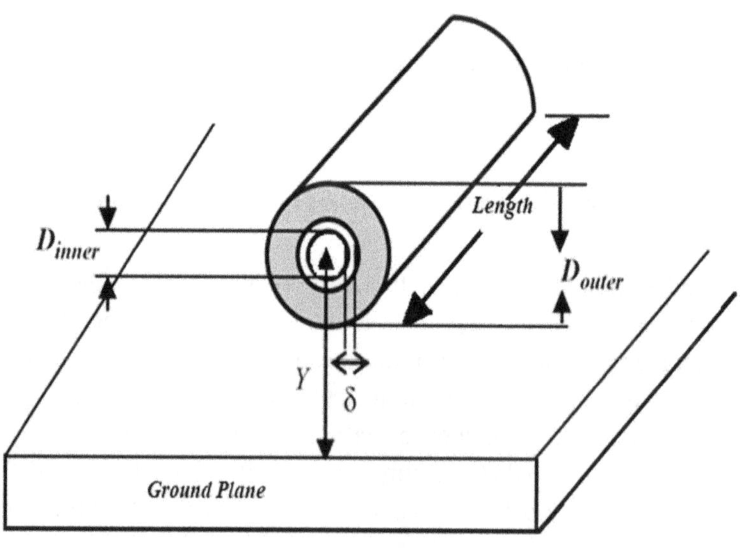

where int {.} represents the integer part which is to be considered for evaluation. Each shell of MWCNT bundle have different diameter, hence the diameter of any ith shell of a bundle can be calculated as (Singh & Raj, 2015; Litoria, Sandha & Kansal, 2017)

$$D_i = D_{outer} - 2\delta\left(i-1\right), \quad for 1 \le i \le n \tag{15.19}$$

Each shell of MWCNT bundle is having different number of conducting channels and these conducting channels are diameter dependent. If the diameter of any ith shell is less than 3nm, then the number of conducting channel will be considered as 2 and for shells having diameters more than 3nm is diameter dependent and can be calculated as (Singh & Raj 2015)

$$N_{\frac{chan}{i^{th} shell}} \approx 2, \qquad for\ D_i < 3nm \tag{15.20}$$

$$N_{\frac{chan}{i^{th} shell}} \approx a.D_i + b, \qquad for \quad D_i > 3nm \tag{15.21}$$

Where Di is the diameter of ith shell, a=6.12x 10^{-4} 1/(nmK) and b=1.275. The total number of conducting channels (Nchan/bundle) in a bundle is the sum of all the conducting channels ($N_{\frac{chan}{shell}}$) of all the shells and given by

$$N_{chan/bundle} = \sum_{i=1}^{n} N_{chan} \ / \ i^{th} shell \tag{15.22}$$

5.2. Impedance Parameters of an Individual Shell of MWCNT Bundle

Based on interconnect parasitic of an individual shell (resistance, capacitance and inductance), the equivalent impedance model for an individual shell of MWCNT is shown in Fig. 15.7.

Resistance (R_{shell}) of ith Shell

The resistance of any i^{th} shell of MWCNT bundle is divided into three parts: scattering dependent resistance (R_s) (if l < λ), quantum resistance (R_q) and metal

Figure 7. Impedance parameter model of ith shell of MWCNT

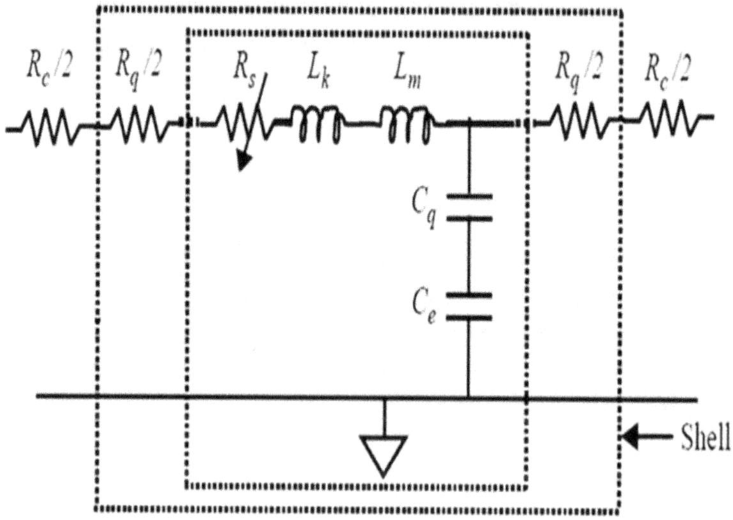

contact resistance (R_c). The metal contact resistance (R_c) is due to the contact of metal electrodes with MWCNT and depends upon the fabricating process. Therefore, metal contact resistance (R_c) is almost constant and assumed equal to *2kΩ* for each shell of MWCNT bundle as suggested in (Singh & Raj, 2015). The total resistance for an individual shell is given by

$$R_{shell} = R_c + R_q + R_s.L$$
$$R_{shell} = 2k\Omega + \frac{h}{2e^2N} + \frac{h}{2e^2N}.\frac{L}{\lambda} \tag{15.23}$$

where $h/2e^2 = 12.9k\Omega$, N, L and λ are the number of conducting channels in a shell, interconnect length and MFP of a shell of MWCNT respectively. (Singh & Raj, 2015; Sandha & Sharma, 2018).

Inductance of ith Shell

There are two types of inductances in MWCNT bundle i.e. the magnetic inductance and kinetic inductance per unit length]. The magnetic inductance is due to the current flowing through MWCNT shell and stored as total magnetic energy in the shell. The magnetic inductance for an individual 'i^{th}'shell is given by

$$L_m = \left(\frac{\mu}{2\pi}\right).cosh^{-1}\left(\frac{2Y}{D_i}\right) \tag{15.24}$$

where μ is mobility of electron, Y is the distance between center of the bundle and ground and D_i is diameter of ith shell as shown in Figure 6. The kinetic inductance is due to each conducting channel of MWCNT shell stored in the form of kinetic energy when shell conducts and given as (Singh & Raj, 2015)

$$L_{k/channel} = \frac{h}{2 \times 2v_f e^2} \tag{15.25}$$

where v_f is Fermi velocity for carbon nanotubes is usually considered as 8x 10^5 m/s, e is electron charge and h is Planck constant. All individual shells and their conducting channels are considered to be in parallel to each other. Hence, the effective kinetic inductance of an individual shell can be calculated by

$$L_{k/shell} = L_{k/channel} / N_i \tag{15.26}$$

The mutual inductance in a bundle is due to the coupling between different shells of MWCNT bundle. The shells of MWCNT bundle have large diameters with small thickness, so each shell can be considered as ideal cylinder with zero thickness. Therefore, the mutual inductance between shells can be estimated around *2pH/μm.*

Capacitance of ith Shell

The capacitance in MWCNT bundle is of two types: quantum capacitance (C_q) and electrostatic capacitance (C_e). The capacitance due to quantum electrostatic charge stored in the nanotube when it carries current is known as quantum capacitance (C_q) (Singh & Raj, 2015). The quantum capacitance for a shell is given by

$$C_{q/channel} = \frac{2 \times 2e^2}{hv_f} \approx 193aF / \mu m \tag{15.27}$$

$$C_{q/shell} = C_{q/channel}.N_i \tag{15.28}$$

The electrostatic capacitance (C_e) appear due to charge stored by the outermost shell of MWCNT bundle with diameter D_{outer} and placed above ground at distance Y, as shown in Figure 6. Hence, the electrostatic capacitance per unit length of shell is given by

$$C_e = \frac{2\pi\varepsilon}{cosh^{-1}(\frac{2Y}{D_{outer}})} \tag{15.29}$$

The potentials of difference of MWCNT shells cannot be considered to be at same level, which introduces inter-shell coupling capacitance. The inter-shell capacitance (Cs) can be obtained by using the same equation which can be used for coaxial capacitance and given by

$$C_s = \frac{2\pi\varepsilon}{ln(\frac{D_{out}}{D_{in}})} = \frac{2\pi\varepsilon}{ln[D_{out} / (D_{out} - 2\delta)]} \tag{15.30}$$

where D_{in} and D_{out} are the diameters for inner and outer adjacent shells of MWCNT bundle respectively and δ=0.34nm.

5.3. Equivalent Circuit Model of MWCNT Bundle Interconnects

An individual shell of MWCNT has very high resistance hence it cannot be used as individual tube for interconnects. So, concentrically nested several MWCNTs are used in parallel to form MWCNT bundle resulting in net reduction of overall resistance. Based on the above mentioned parameters, equivalent impedance model for MWCNT bundle interconnects is shown in Figure 8 in which all the shells of bundle are considered as parallel shells. (Singh & Raj, 2015; Sandha & Sharma, 2018).

6. CIRCUIT MODELING OF COPPER INTERCONNECTS

The impedance equivalent model for copper as interconnects material is presented in this section. The schematic structure used to evaluate equivalent impedance parameters such as resistance, inductance and capacitance for a copper interconnect is shown in Figure 9.

Figure 8. Equivalent impedance model for a MWCNT bundle as interconnects

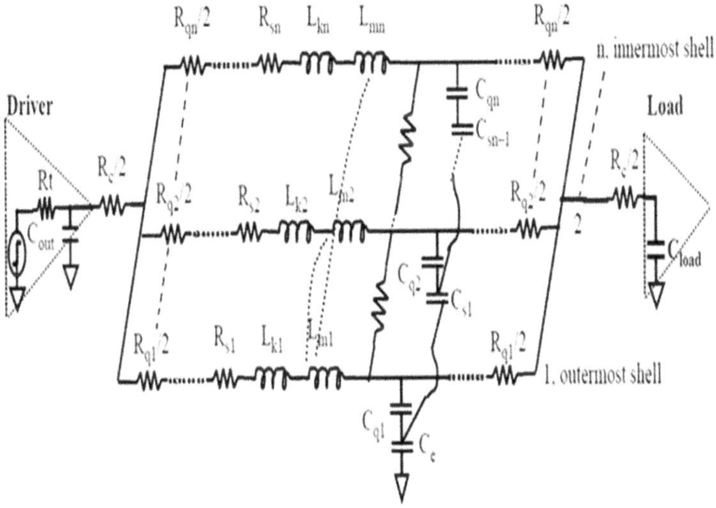

Figure 9. Interconnect geometry for copper

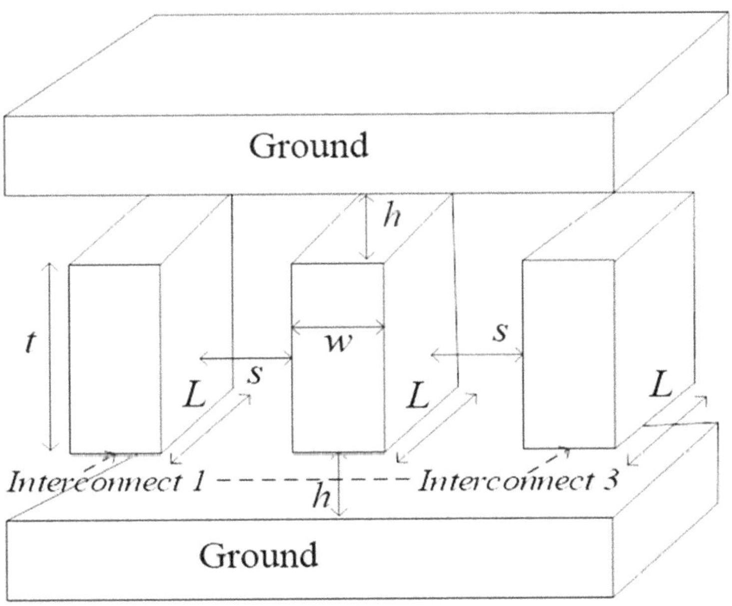

6.1. Resistance

The resistivity of copper plays an important role to calculate the resistance for rectangular cross-sections interconnects as shown in Figure 9. Therefore, the resistance of copper interconnect depends upon the resistivity and technology dependent cross-sectional dimensions as given below

$$R_{Cu} = \frac{pL'pp+}{W_t W_t} \tag{15.31}$$

where, ρ is resistivity of copper and resistivity ρ divided into two parts and given as phonon surface scattering (ρ_s) and defect (ρ_d) and given as

$$\rho = \rho_s + \rho_d \tag{15.32}$$

$$\rho = \rho_0 \left[1 + 0.00401\left(T - 300\right)\right] \tag{15.33}$$

where ρ0 is technology dependent resistivity of copper at 300K. (Pop et al., 2007; Singh & Raj, 2015)

6.2. Inductance and Capacitance

The inductance and capacitance of copper interconnects can be expressed as

$$L_{Cu} = \frac{\mu_0 L}{2\pi}\left[\ln\left(\frac{2L}{W+t}\right) + \frac{1}{2} + \frac{0.22(W+t)}{L}\right] \tag{15.34}$$

$$C_g = \varepsilon\left[\frac{W}{h} + 2.22\left(\frac{s}{s+0.7h}\right)^{3.19} + 1.17\left(\frac{s}{s+1.51h}\right)^{0.76} \cdot \left(\frac{t}{t+4.53h}\right)^{0.12}\right] \tag{15.35}$$

Where, Cg is capacitance of per unit length, W, h, t are technology dependent the cross-sectional dimensions of Copper interconnects as shown in Figure 9.

7. RESULTS AND DISCUSSION

All the impedance parameters of different materials have been calculated by using the analytical models presented in the different sections of the chapter. The parasitic of SWCNT, MWCNT and copper have been evaluated by using equations (15.1-15.17), (15.18-15.30) and (15.31-15.35) respectively at 32, 22 and 16nm technology nodes for 500, 1000, 1500 and 2000μm interconnect length. The MATLAB is used to write the codes for different equations to calculate the impedance parameters of different materials. The interconnect parameters used to calculate the parasitic with the help of equations for different material are taken from ITRS 2013 (http://public.itrs.net) and shown in Table 3

The calculated resistances for different materials at 32, 22 and 16 nm technology nodes for 500, 1000, 1500 and 2000μm length are shown in Table 4.

It can be well interpreted from the values shown in Table 4 that for 32nm technology node and below, because of certain issues prevailing in copper like electromigration and other scattering effects as mentioned in the previous sections, there is a sharp rise in its value of resistance and this value gets significantly large as technology nodes are scaled down from 32nm to 16nm. The CNTs (SWCNT and MWCNT) show the lesser value of resistance and are introduced because of their extraordinary good conduction properties making them highly applicable as VLSI interconnect material.

For the comparative analysis, the evaluated results for 1000μm interconnects length are plotted in Figure 10 for better understanding of the results. It is revealed from the results shown in Figure 10 that the gap between copper and CNTs is increasing with scaled down technology nodes i.e. 32nm to 16nm, which is due to the scattering mechanism effects and electromigration.

Table 3. Interconnect parameters for global level interconnect length as per ITRS 2013 (http://public.itrs.net)

Interconnect Parameters	Technology Node		
	32nm	**22nm**	**16nm**
Width of interconnect	40nm	28nm	18nm
Height of interconnect	120nm	84nm	54nm
Thickness of Oxide	93.6nm	65.5nm	40nm
Aspect ratio (A/R)	3	3	3
Vdd (volts)	0.9	0.8	0.7
Dielectric constant (ε)	2.77	2.59	2.31
Resistivity for copper (μΩ-cm)	3.66 μΩ-cm	4.2 μΩ-cm	5.69 μΩ-cm

Table 4. Comparative analysis in term of resistance between SWCNT, MWCNT and copper interconnects for interconnect length 500, 1000, 1500 and 2000µm at 32nm, 22nm and 16nm technologies

Interconnect Length	Technology Nodes								
	32 nm			22 nm			16 nm		
	Copper	SWCNT	MWCNT	Copper	SWCNT	MWCNT	Copper	SWCNT	MWCNT
500um	3812.5	674.7681	1043.599	8928.571	1377.078	2795.354	29269.55	3332.188	7115.448
1000um	7625	1348.189	2074.283	17857.14	2751.407	5556.116	58539.09	6657.725	14142.84
1500um	11437.5	2021.611	3104.965	26785.71	4125.736	8316.871	87808.64	9983.262	21170.22
2000um	15250	2695.032	4135.647	35714.29	5500.065	11077.62	117078.2	13308.8	28197.59

Figure 10. Comparison of resistance between Copper, SWCNT and MWCNT interconnects for 1000µm length at 32, 22 and 16nm technologies

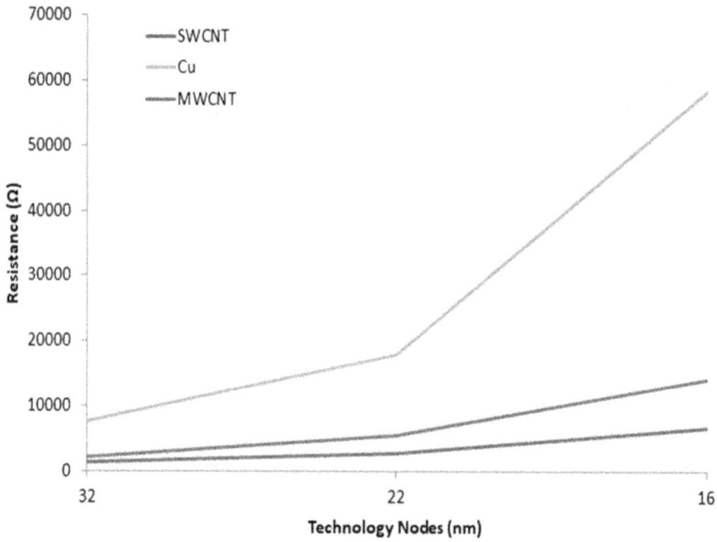

The calculated parasitic are used to simulate the performance in terms of delay and PDP. The HSPICE simulations tool is used for Copper, SWCNT and MWCNT at interconnect length 500µm to 2000µm at different nanometer technology nodes to study their propagation delays. (www.ptm.asu.edu) The simulation results for copper, SWCNT and MWCNT as interconnects material for interconnect length 500 µm-2000 µm are shown in Figure 11 (a)- (c) at 32 nm, 22nm and 16 nm technology nodes respectively.

Figure 11a. Comparison of delay between copper, SWCNT and MWCNT interconnects length from 500μm- 2000μm at 32nm technologies respectively

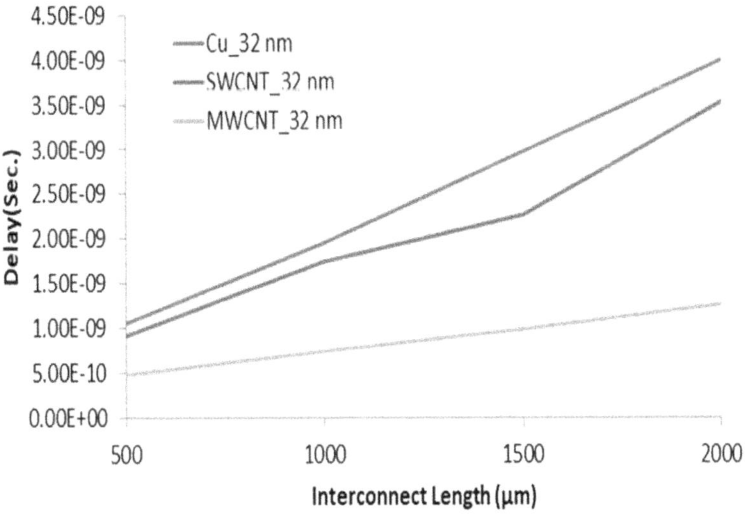

Figure 11b. Comparison of delay between copper, SWCNT and MWCNT interconnects length from 500μm- 2000μm at 22nm technologies respectively

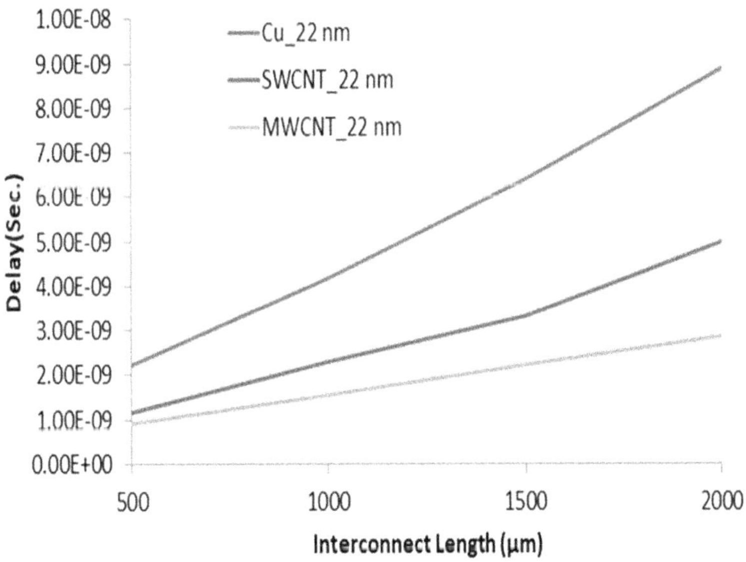

Figure 11c. Comparison of delay between copper, SWCNT and MWCNT interconnects length from 500µm- 2000µm at 16nm technologies respectively

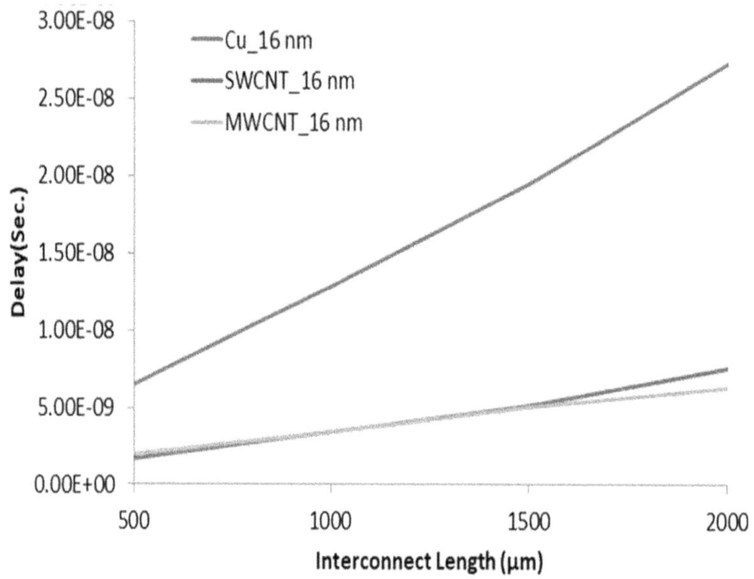

It is clearly depicted from the results shown in Figure 11 (a)-(c) that delay of circuit rises with the increase in the length of interconnects. As observed from the values obtained, carbon nanotubes show the lesser value of propagation delay than copper which suffers from electromigration and surface scattering effects which leads to increased resistivity and decreased mean free path at global lengths at scaled future technology nodes.

Further, this paper presents graphs of performance parameter and figure of merit i.e. PDP (power delay product) of copper, SWCNT and MWCNT for 500 µm to 2000 µm shown in Figure 12 (a)-(c) at 32nm, 22nm and 16nm technology nodes respectively.

It is clearly evident from the results shown in Figure 12 (a)-(c) that the value of the PDP parameter of CNT is again lowest as compared to copper. As observed from the results that with the further scaling of technology node, the ratio of propagation delay and PDP of both types of CNT with respect to copper shows an improved performance in favor of former which is due to decreased MFP and fatal issues faced by copper at nanometer technology nodes for global lengths of interconnect. Clearly, CNT can outperform the existing interconnect materials for next generation VLSI technology nodes.

Figure 12a. Comparison of PDP between copper, SWCNT and MWCNT interconnects for interconnect length 500µm-2000µm length at 32nm technology nodes

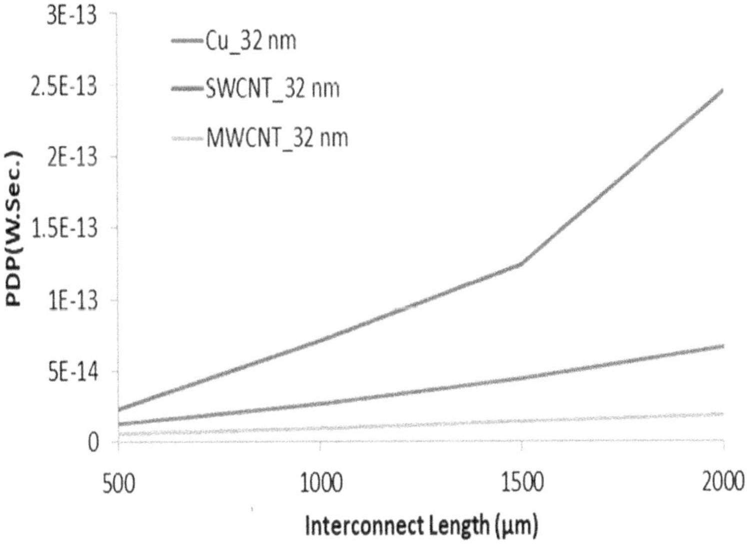

Figure 12b. Comparison of PDP between copper, SWCNT and MWCNT interconnects for interconnect length 500µm-2000µm length at 22nm technology nodes

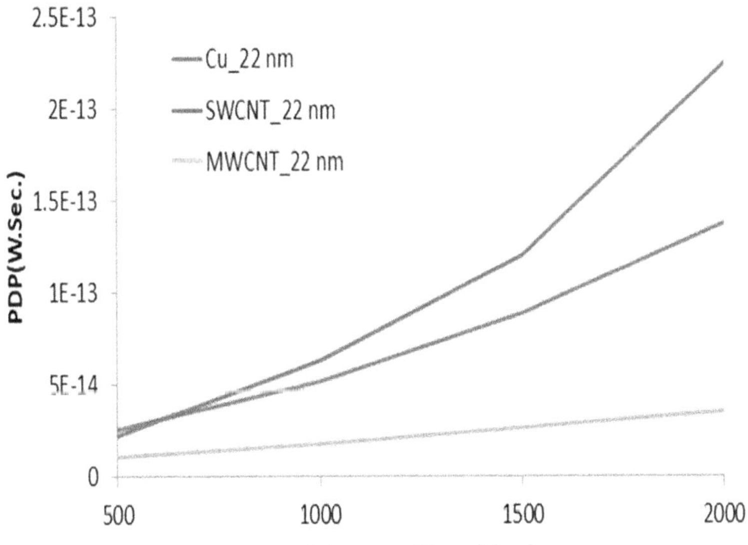

Figure 12c. Comparison of PDP between copper, SWCNT and MWCNT interconnects for interconnect length 500μm-2000μm length at 16nm technology nodes

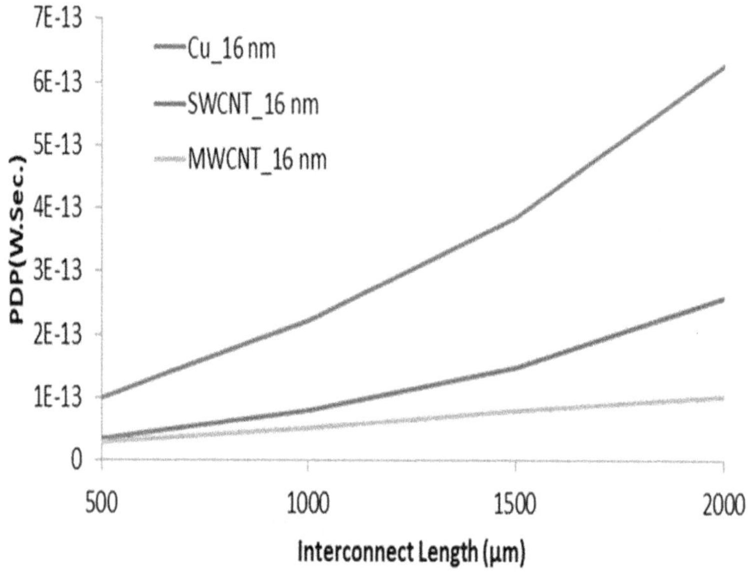

7. CONCLUSION AND FUTURE ASPECTS

This chapter presented the deficiencies of copper as interconnect at nano-scaled technology nodes and advantages of CNTs (both SWCNT and MWCNT) as interconnects are highlighted so that some improvement in the performance could be established. The electrical equivalent circuit models for SWCNT, MWCNT and copper interconnect are also presented. The impedance parameters on the basis of electrical equivalent circuit models are evaluated for different interconnect lengths (500-1000μm) at 32nm, 22nm and 16nm technology nodes. By the virtue of Tanner EDA simulations performed for variable interconnect lengths and results show that the both types of CNT i.e. single-wall, multi-wall and multi-wall CNT undoubtedly show better performance than copper in terms of propagation delay and power-delay product which suffers from several issues like decreased MFP and surface scattering effects at nanometer technology nodes and this improvement in results gets better with the further scaling of technology nodes.

The fabrications of CNT based interconnects have certain real challenges for its growing process. Therefore, the issues related for fabrication process of MWCNT is needs to be taken care of in future work. Further, in the recent studies, mixed CNT interconnects (SWCNT and MWCNT) are also proposed for VLSI applications. Therefore, accurate analysis of Mixed CNT based interconnects need to be estimated.

REFERENCES

Bakoglu, H. B., & Meindl, J. D. (1985). Optimal interconnection circuits for VLSI. *IEEE Transactions on Electron Devices*, *32*(5), 903–909. doi:10.1109/T-ED.1985.22046

Bourlon, B., Miko, C., Forro, L., Glattli, D. C., & Bachtold, A. (2004). Determination of the intershell conductance in multiwalled carbon nanotubes. *Physical Review Letters*, *93*(17), 176806. doi:10.1103/PhysRevLett.93.176806 PMID:15525107

Burke, P. J. (2002). Luttinger liquid theory as a model of the gigahertz electrical properties of carbon nanotubes. *IEEE Transactions on Nanotechnology*, *1*(3), 129–144. doi:10.1109/TNANO.2002.806823

Das, P. K., Majumder, M. K., & Kaushik, B. K. (2014). Dynamic crosstalk analysis of mixed multi-walled carbon nanotube bundle interconnects. *The Journal of Engineering, 2014*(5), 227-233.

Gholipour, M., & Masoumi, N. (2012). Efficient inclusive analytical model for delay estimation of multi-walled carbon nanotube interconnects. *IET Circuits, Devices & Systems*, *6*(4), 252–259. doi:10.1049/iet-cds.2011.0283

Hosseini, A., & Shabro, V. (2010). Thermally-aware modeling and performance evaluation for single-walled carbon nanotube-based interconnects for future high performance integrated circuits. *Microelectronic Engineering*, *87*(10), 1955–1962. doi:10.1016/j.mee.2009.12.008

Im, S., Srivastava, N., Banerjee, K., & Goodson, K. E. (2005). Scaling analysis of multilevel interconnect temperatures for high-performance ICs. *IEEE Transactions on Electron Devices*, *52*(12), 2710–2719. doi:10.1109/TED.2005.859612

International Technology Roadmap for semiconductor. (2013). Retrieved from http://public.itrs.net

Kaushik, B. K., & Majumder, M. K. (2015). Carbon nanotube: Properties and applications. In *Carbon Nanotube Based VLSI Interconnects* (pp. 17–37). New Delhi: Springer.

Li, H., Lu, W. G., Li, J. J., Bai, X. D., & Gu, C. Z. (2005). Multichannel ballistic transport in multiwall carbon nanotubes. *Physical Review Letters*, *95*(8), 086601. doi:10.1103/PhysRevLett.95.086601 PMID:16196881

Li, H., Xu, C., Srivastava, N., & Banerjee, K. (2009). Carbon nanomaterials for next-generation interconnects and passives: Physics, status, and prospects. *IEEE Transactions on Electron Devices*, *56*(9), 1799–1821. doi:10.1109/TED.2009.2026524

Li, H., Yin, W. Y., Banerjee, K., & Mao, J. F. (2008). Circuit modeling and performance analysis of multi-walled carbon nanotube interconnects. *IEEE Transactions on Electron Devices*, *55*(6), 1328–1337. doi:10.1109/TED.2008.922855

Litoria, P., Sandha, K. S., & Kansal, A. (2017). Impact of tunneling conductance on the performance of multi walled carbon nanotubes as VLSI interconnects for nano-scaled technology nodes. *Journal of Materials Science Materials in Electronics*, *28*(6), 4818–4827. doi:10.100710854-016-6128-3

Majumder, M. K., Kaushik, B. K., & Manhas, S. K. (2014). Analysis of delay and dynamic crosstalk in bundled carbon nanotube interconnects. *IEEE Transactions on Electromagnetic Compatibility*, *56*(6), 1666–1673. doi:10.1109/TEMC.2014.2318017

McEuen, P. L., Fuhrer, M. S., & Park, H. (2002). Single-walled carbon nanotube electronics. *IEEE Transactions on Nanotechnology*, *1*(1), 78–85. doi:10.1109/TNANO.2002.1005429

Naeemi, A., & Meindl, J. D. (2006). Compact physical models for multiwall carbon-nanotube interconnects. *IEEE Electron Device Letters*, *27*(5), 338–340. doi:10.1109/LED.2006.873765

Naeemi, A., & Meindl, J. D. (2009). Carbon nanotube interconnects. *Annual Review of Materials Research*, *39*(1), 255–275. doi:10.1146/annurev-matsci-082908-145247

Pop, E., Mann, D., Reifenberg, J., Goodson, K., & Dai, H. (2005, December). Electro-thermal transport in metallic single-wall carbon nanotubes for interconnect applications. In *IEEE InternationalElectron Devices Meeting, 2005. IEDM Technical Digest.* (pp. 4-pp). IEEE. Retrieved from www.ptm.asu.edu

Pop, E., Mann, D. A., Goodson, K. E., & Dai, H. (2007). Electrical and thermal transport in metallic single-wall carbon nanotubes on insulating substrates. *Journal of Applied Physics*, *101*(9), 093710. doi:10.1063/1.2717855

Rossnagel, S. M., & Kuan, T. S. (2004). Alteration of Cu conductivity in the size effect regime. *Journal of Vacuum Science & Technology. B, Microelectronics and Nanometer Structures: Processing, Measurement, and Phenomena: An Official Journal of the American Vacuum Society*, *22*(1), 240–247.

Rutherglen, C., & Burke, P. (2009). Nanoelectromagnetics: Circuit and electromagnetic properties of carbon nanotubes. *Small*, *5*(8), 884-906.

Sandha, K. S., & Sharma, S. (2018). Performance Analysis of Different Mixed-MWCNT Structures as VLSI Interconnects for Nano-Electronics IC Design. *Journal of Nanoelectronics and Optoelectronics*, *13*(3), 357–367. doi:10.1166/jno.2018.2229

Singh, K., & Raj, B. (2015). Influence of temperature on MWCNT bundle, SWCNT bundle and copper interconnects for nanoscaled technology nodes. *Journal of Materials Science Materials in Electronics*, *26*(8), 6134–6142. doi:10.100710854-015-3193-y

Singh, K., & Raj, B. (2015). Performance and analysis of temperature dependent multi-walled carbon nanotubes as global interconnects at different technology nodes. *Journal of Computational Electronics*, *14*(2), 469–476. doi:10.100710825-015-0667-3

Singh, K., & Raj, B. (2015). Temperature-dependent modeling and performance evaluation of multi-walled CNT and single-walled CNT as global interconnects. *Journal of Electronic Materials*, *44*(12), 4825–4835. doi:10.100711664-015-4040-x

Srivastava, N., Li, H., Kreupl, F., & Banerjee, K. (2009). On the applicability of single-walled carbon nanotubes as VLSI interconnects. *IEEE Transactions on Nanotechnology*, *8*(4), 542–559. doi:10.1109/TNANO.2009.2013945

Steinhögl, W., Schindler, G., Steinlesberger, G., Traving, M., & Engelhardt, M. (2005). Comprehensive study of the resistivity of copper wires with lateral dimensions of 100 nm and smaller. *Journal of Applied Physics*, *97*(2), 023706. doi:10.1063/1.1834982

Chapter 8
Bundled SWCNT for Global VLSI Interconnects

Raj Kumar
Panjab University, India

Shashi Bala
Chandigarh Group of Colleges, Landran, India

ABSTRACT

Carbon nanotube (CNT) has been declared the most attractive and suitable material for VLSI sub-micron technology. Because of CNT's phenomenal physical, electrical, and mechanical properties, it is more advantageous than copper interconnect material. In this chapter, RLC equivalent model of bundled single-wall CNT (SWCNT) is presented by using driver-interconnect-load (DIL) system with CMOS driver. The crosstalk delay is calculated for two-line bus architecture made of two parallel lines (i.e., upper as aggressor and lower as victim). From the simulation, it has been observed that crosstalk delay increases with increase in interconnect length and transition time, whereas it decreases with increased spacing between the lines (aggressor and victim). However, crosstalk delay decreases as the number of tubes in a bundle increases.

1. INTRODUCTION

Interconnects are basically used to establish communication between various components on-chip and modules on system board. Interconnects has become important to take into account in deep submicron technology. As the technology shrinking, the minimum width of 5nm interconnects and the minimum spacing

DOI: 10.4018/978-1-7998-1393-4.ch008

between two interconnects decrease. It is When looking at the scaling of on-chip interconnects, it is important to study interconnects because in deep sub-micron technology interconnects are no longer remain as resistor (Banerjee & Srivastava, 2006). The related parasitic such as inductance and capacitance are also important to take into consideration in deep submicron technologies. These parasitic are responsible for delay in a propagating signal through interconnects (Wei, Vajtai, & Ajayan, 2001; Srivastava, Joshi & Banerjee, 2005). As the size of the chip increases the interconnects also come very close to each other thus the coupling capacitance occurs between the wires or interconnects will increases (Naeemi & Meindl, 2006; Rossi, Cazeaux, Metra & Lombardi, 2007). The feature size of ICs is shrinking to improve the propagation delay, power consumption, silicon area, and cost characteristics. The technology scaling in VLSI interconnects causes increase in density. This increased density results in the increasing of the coupling effect. Interconnect plays a vital role in the performance of a high-speed VLSI chip.

Earlier, metals were used as VLSI interconnects material. In the last few years, carbon nanotubes (CNTs) have been appeared as most attractive and suitable interconnects material for future deep sub-micron technology. Carbon nanotubes have phenomenal current carrying capability, high thermal conductivity, high ballistic transport, resilient to electromigration and tremendous mechanical properties. Carbon nanotubes are considered for VLSI interconnects because of these remarkable properties (Srivastava, Joshi, & Banerjee, 2005; Srivastava & Banerjee, 2005).

Crosstalk coupling effect affects the righteousness of signal by producing glitches and unwanted voltage spikes on the adjacent lines. It also causes a delay at the output port of the system. The delay appeared at output mainly rely on upon coupling capacitance and relative strength of drivers, that drives the signal into the lines either in-phase or out-of-phase, that is responsible for the different skew rates of signal to the components. Therefore, reliability has been a major concern of crosstalk effect. Crosstalk noise effect in coupled lines splits into two types: (1) functional crosstalk noise (2) dynamic crosstalk noise (Javey & Kong, 2009; Pu, Yin, Mao, & Liu, 2009; Majumder, Das, Kaushik, & Manhas, 2012). If the victim line undergoes a voltage spike when signal applied to an aggressor line is presumed under functional crosstalk category (Majumder, Das, Kaushik, & Manhas, 2012). Whereas, dynamic crosstalk is calculated as the signal applied to aggressor and victim line simultaneously. This paper emphasized on dynamic crosstalk effect. The crosstalk mainly causes by parasitics such as resistance, inductance and capacitance. The crosstalk effect will decrease with reduction in parasitic. In this paper, numbers of SWCNTs are increased in a bundle to reduce the value of parasitics. Thereby, the crosstalk delay was analyzed for two bus line architecture.

2. TYPES OF INTERCONNECTS MATERIAL

2.1 Copper Interconnects

Semiconductor integrated circuits basically use copper wires for interconnects, which are mostly microprocessors. Copper is considered as a better conductor of electricity than aluminum, chips that are using this technology could have smaller metal component. In result use very less energy to pass electricity through interconnects. Ultimately, these effects are responsible for higher-performance processors. Earlier, copper was used as VLSI interconnect

material. But due to its less resistance to electromigration, designer had to move towards CNTs as VLSI interconnects material (Majumder, Das, Kaushik, & Manhas, 2012).

2.2 Optical Interconnects

Optical interconnects are used for communication through optical cables. Optical wires have higher bandwidth capacity of 10 Gbits/sec to 100 Gbits/sec. Optical interconnects have a negligible frequency-dependent loss, low cross talk, and high bandwidth. Chip-to-chip optical interconnections operate at much higher speeds allowing the multiplexing of a large number of I/O signals in a single I/O fiber. Optical interconnects are not generally used for commercial purpose. Optical interconnects currently are not used in the semiconductor industry. Because optical interconnects technology is not compatible with manufacturing processes and methods used to assemble.

2.3 Carbon Nanotube Interconnects

Carbon nanotubes are basically allotropes of carbon with the cylindrical-shaped nanostructure. Nanotubes have been generally made with length-to-diameter ratio of up to 132,000,000:1. It has remarkable thermal conductivity and mechanical and electrical properties. Nanotubes are the member of the fullerene family. Carbon nanotubes are classified as single-walled nanotubes (SWCNTs) and multiwalled nanotubes (MWCNTs). On the basis of Chirality CNT are classified as armchair, zigzag and chiral.because of its phenomenal properties CNTs preferred over copper interconnects ((Majumder, Das, Kaushik, & Manhas, 2012; Mittal & Lin, 20170.

3. CARBON NANOTUBES (CNT)

In the last few years, CNTs have emerged as promising interconnects material for future deep sub-micron technologies. CNTs are constructed of sheets of graphene rolled up as hollow cylinders. Carbon nanotubes are very strong in nature because of properties of graphene. Carbon nanotubes have a phenomenal current carrying capability, high thermal conductivity, high ballistic transport, and tremendous mechanical properties. Carbon nanotubes are considered for VLSI interconnects because of these remarkable properties (Srivastava & Banerjee, 2005).

3.1 Types of Carbon Nanotubes (CNTs)

CNTs are classified into two categories as explained below.

3.1.1 Single-Walled Carbon Nanotubes (SWCNTs)

Single-walled carbon nanotubes (SWCNTs) are made up single graphene layers rolled as hollow cylinder as shown in figure 1. The diameter of SWCNTs is generally ranging from 0.4 to 4 nm. SWNTs are most important variety of carbon nanotube. The bandgap of SWCNTs varies from 0 to 2 eV. Single-walled carbon nanotubes are better candidate for VLSI technology. Single-walled carbon nanotubes are very expensive material. One useful application of SWNTs is in the development of the first intermolecular field-effect transistors (FET) (Das & Rahaman, 2012; Majumder, Kaushik & Manhas, 2014).

Figure 1. Single-Walled Carbon Nanotubes (SWCNTs)

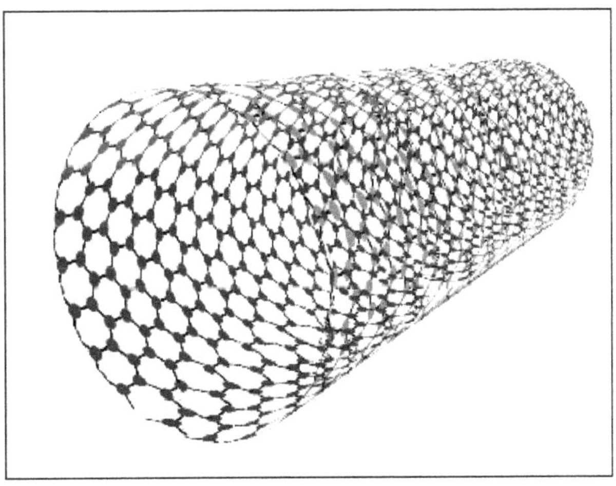

3.1.2 Multi-Walled Carbon Nanotubes (MWCNTs)

Multiwalled carbon nanotubes (MWCNTs) are made by a number of concentric shells or graphene sheets rolled over again and again, as shown in fig 2. The diameter of MWCNTs generally ranging from few nanometers to tens of nanometers. There are two ways to design MWCNTs:

1. **Russian Doll Model**: In this model, sheets of graphite are arranged in concentric cylinders.
2. **Parchment Model**: A single sheet of graphite is rolled in around itself like a newspaper.

The interlayer distance in multi-walled nanotubes is close to the distance between graphene layers in graphite, approximately 3.4 Å. The Russian Doll structure is mostly used. The individual shells of MWCNTs are SWCNTs (Majumder, Kaushik, Manhas, & Das, 2012; Majumder, Pandya, Kaushik, & Manhas, 2012; Kumar, Chandel, & Agrawal, 2017).

3.2 Chirality

Nature of carbon nanotubes (CNTs) depends upon its chiral indices like the direction in which they get rolled up. SWCNTs may be either metallic or semiconducting in nature depends on their chiral indices (n,m). The integers n and m denote the number of unit vectors along two directions in the honeycomb crystal lattice of graphene.

Figure 2. Multi-Walled Carbon Nanotubes (MWCNTs)

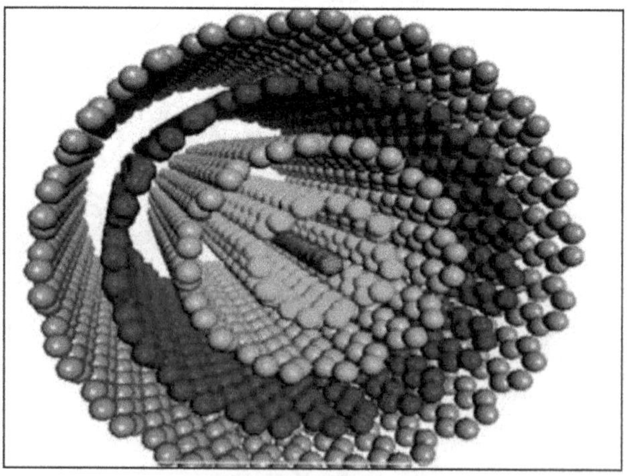

The structures of carbon nanotube depend on chiral indices as shown in Figure 3. Zigzag CNTs made when n or m=0 defined. Zigzag CNTs are either semiconducting or metallic in nature shown in Figure 4 (Das & Rahaman, 2012; Majumder, Pandya, Kaushik & Manhas, 2012).

When n=m. then Armchair CNTs formed. Armchair CNTs are mostly metallic in nature. The structure of Armchair CNT's shown in Figure 5 below.

4. GLOBAL VLSI INTERCONNECTS

As per Moore's law density of transistor increases on single-chip over 30 years, the feature size of CMOS technology is also shrinking to dimensions into the nanometer region nowadays. As a result of this continuous scaling, higher circuit speeds, lower power and larger packing densities of transistors are achieved. Both the thickness of the metal layers itself and the thickness of the oxide between the metal layers decrease with scaling. Interconnects basically used to establish communication between various components on-chip and modules on system board. Interconnects has become important to take into account in deep submicron technology. As the technology shrinking, the minimum width of an interconnect and the minimum spacing between two interconnects decrease. When looking at the scaling of on-chip interconnects, it is important to distinguish between local and global interconnects.

Figure 3. Honeycomb crystal lattice

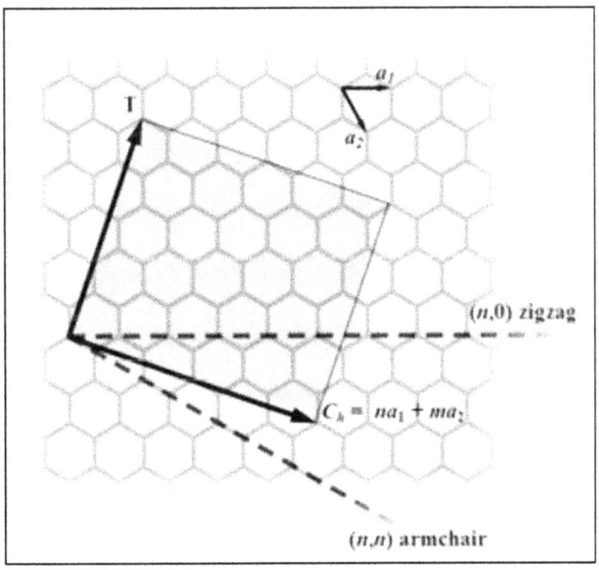

Figure 4. Zigzag CNT

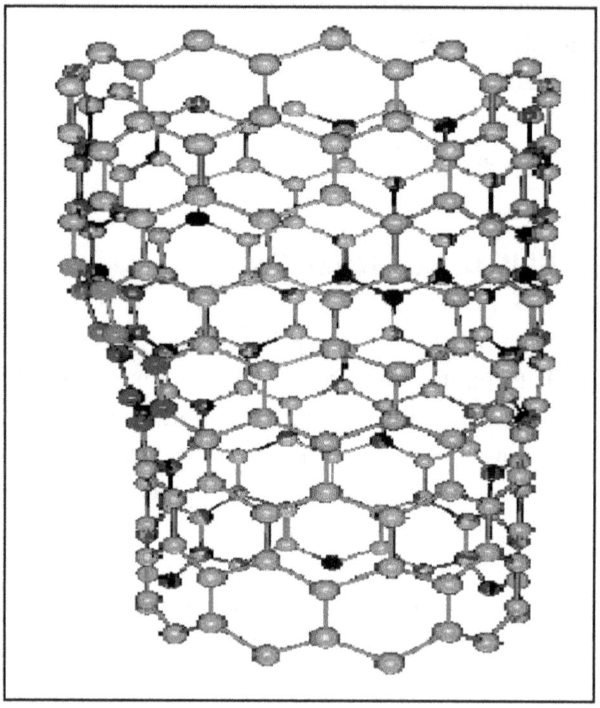

Figure 5. Armchair CNT

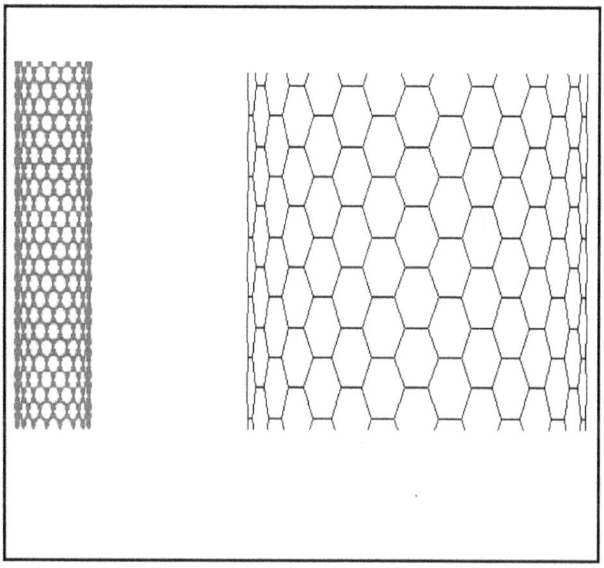

Figure 6 is representing the structure of interconnects. There are modules present on system board, the wires shown inside the modules are known as local level interconnects, whereas wires shown connecting two or more than two modules are labeled as global interconnects. Copper can be used for local level interconnects. But for global level interconnects CNTs are strictly preferred in deep sub-micron technology (Naeemi & Meindl, 2006; Uhlig et al., 2018).

5. APPLICATIONS

As the technology and feature size shrinking with such a high pace, reliability and performance of the chip much depend on the interconnects. Nowadays, the gate delay has dominated by interconnect delay. Therefore, to have better performance CNTs are considered as future VLSI interconnects.

Replacement for Copper: At the global level, conventional interconnect material Cu and Al are very much affected by electromigration problems and their current carrying capacity is very low as compared to CNTs. CNTs bundles are used where requires high transmission efficiency and low reflection losses. The numbers of conducting channels in SWCNTs bundle are more than the number of conducting channels in MWCNT per shell of the same size. So it is concluded that MWCNT and SWCNT bundle could be a better option to replace Copper as interconnection wires for next generation of VLSI integrated circuits (Naeemi & Meindl, 2006).

Versatile Applications: In the area of nanoelectronics, CNTs show their chances as energy storage devices (such as supercapacitors), energy conversion devices (including thermoelectric and photovoltaic devices), field emission display sand radiation sources, nanometer semiconductor transistors, nanoelectromechanical systems (NEMS), electrostatic discharge (ESD) protection, as well as interconnects

Figure 6. Types of interconnects

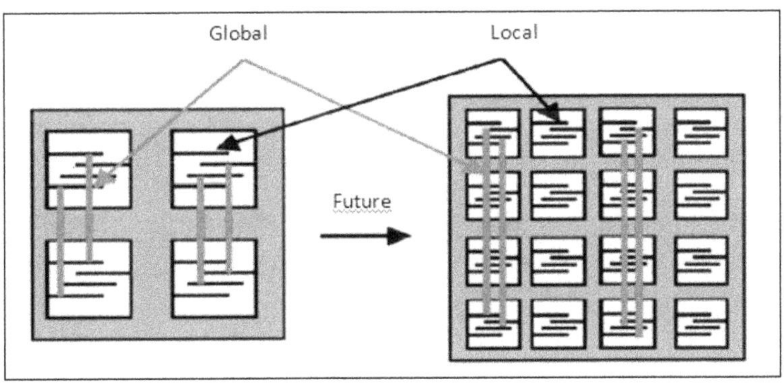

and passives (Banerjee & Srivastava, 2006; Bhatnagar, Merkley, Berdine, Li, & Roy, 2018). The phenomenal physical properties of CNTs make them exciting prospects for a diversity of applications in microelectronics/nanoelectronics, material science, optics, mechanical and biological fields.

6. RLC EQUIVALENT ELECTRICAL CIRCUIT OF SINGLE AND BUNDLED SWCNT

SWCNT and bundled SWCNT are presented in *RLC* electrical models. The geometry of interconnect and its parasitics (resistance, capacitance, and inductance) are described and analyzed here.

6.1 Single SWCNT

SWCNTs are Considered as of two parallel wire model as shown in Figure 7. The diameter of each SWCNT is taken as *d*, the spacing between two SWCNTs is taken as *sp,* distance from the ground plane is as *H* and the length of the SWCNTs labeled as *l* (Kaushik, Sarkar, Agarwal, & Joshi, 2010). According to geometrical structure of SWCNTs, an equivalent *RLC* circuit model of two parallel single SWNTs is shown in Figure 8.

6.1.1. Single SWCNT Resistance

Figure 8 depicts the equivalent electrical model of an isolated single-walled carbon nanotube (SWCNT) for which length of CNTs is taken less than the mean free path of electrons in a CNT (λ_{CNT}), which is considered typically > 1 um. Each carbon

Figure 7. The geometrical structure of SWCNTs in parallel on a ground plane (Kumar, Majumder, Kaushik, Dasgupta, & Manhas, 2012)

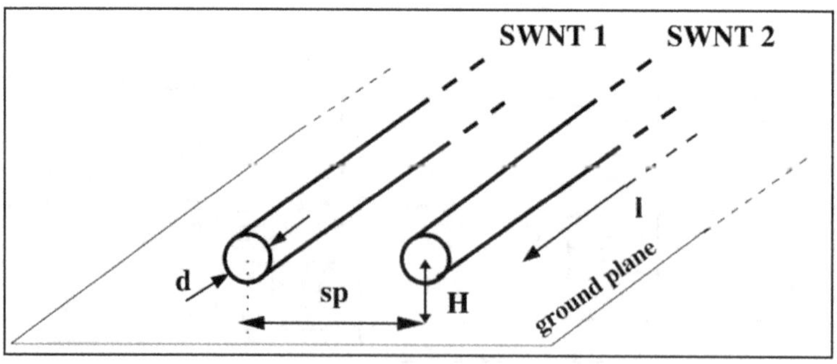

Figure 8. Equivalent RLC electrical circuit model for two-SWCNT

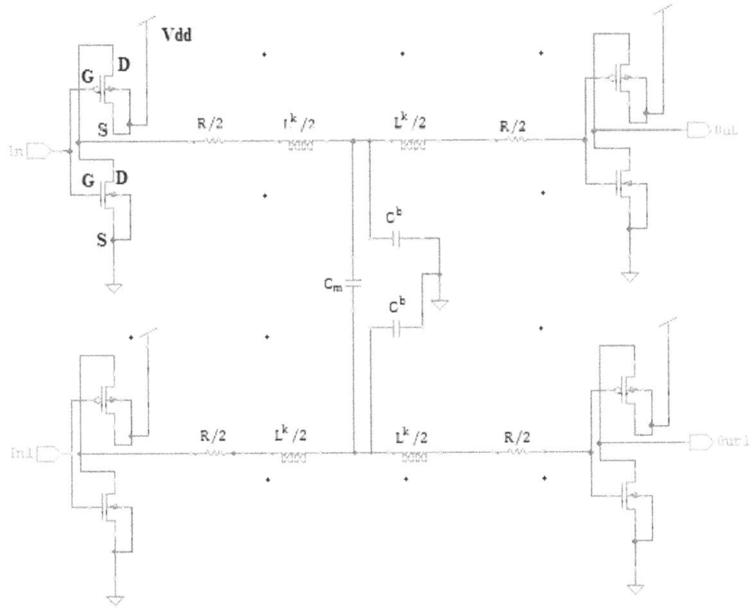

nanotube (CNT) possesses four 1-D conducting channels in parallel due to spin degeneracy and sub-lattice degeneracy of electrons in graphene. The resistance of a carbon nanotube (CNT) for ideal coupling is given below (Kaushik, Sarkar, Agarwal, & Joshi, 2010; Kumar, Majumder, Kaushik, Dasgupta, & Manhas, 2012; Uhlig et al., 2018).

$$R_{CNT} = R_F = \frac{h}{4e^2} = 6.45K\Omega \tag{1}$$

h = Planck's constant and e = electron charge.

If the interconnects length greater than the mean free path i.e. $L > \lambda_{CNT}$, resistance becomes

$$R_S = \left(\frac{h}{4e^2}\right)\frac{L}{\lambda_{CNT}} \tag{2}$$

It gives a per unit length resistance of

$$R_S = \left(\frac{h}{4e^2}\right)\frac{1}{\lambda_{CNT}} = \frac{R_F}{\lambda_{CNT}} \tag{3}$$

Due to metal CNT contact, the imperfect contact resistance (Rc) is taken equals to 100 KΩ and λ_{high} (~26nm) and λ_{low} (~1.6nm) are the mean-free paths (Rossi, Cazeaux, Metra, & Lombardi, 2007). The resistance of SWCNT does not depend at small interconnect length due to the ballistic transport of electrons in CNTs (Loumi, Ghammaz, & Belahrach, 2019). However, resistance increases at long interconnect due to no ballistic transport of electrons in SWCNTs.

6.1.2. CNT Capacitance

The capacitance of a CNT mainly comes from two sources.

The electrostatic capacitance (C_{ES}): The electrostatic capacitance (*CCS* or *CES*) as presents in Fig. 8, the electrostatic coupling between parallel lines and between a CNT and ground plane are given by (Kumar, Majumder, Kaushik, Dasgupta, & Manhas, 2012; Majumder, Kaushik, & Manhas, 2011)

$$C_{ES} = \frac{2\pi \in}{\cosh^{-1}\left(\frac{2H}{d}\right)} \tag{4}$$

$$C_m = \frac{\pi \in}{\cosh^{-1}\left(\frac{sp}{d}\right)} \tag{5}$$

$$C_{ES} = \frac{2\grave{A}\in}{\cosh^{-1}\left(\frac{2H}{d}\right)}$$

The quantum electrostatic energy, that is stored in the carbon nanotube due to which quantum capacitance (C_Q) taken into account when it carries current. Quantum capacitance is used to estimate the energy needed to add an electron to an available quantum state present above the Fermi level. The equation for the quantum capacitance (per unit length) has been described below.

Where h is the Planck's constant and V_F is the Fermi velocity.

$$C_Q = \frac{2e^2}{hV_F} \tag{6}$$

For a carbon nanotubes (V_F= 8x10^5 m/s), C_Q=100 aF/um.

The effective quantum capacitance comes from 4- parallel line capacitances C_Q is given by $4C_Q$ because a CNT has four conducting channels as depicts in the previous section. The effective charge inhabits on both these capacitances (C_E and $4C_Q$) while the CNT carries current. This is true for any two capacitances that are in series. Therefore, they are connected in series as shown in Figure 8.

6.1.3. CNT Inductance

The inductance related with single SWCNT is calculated from the magnetic field of single current carrying wire, which is away from a ground plane, as depicted in Figure 2(b). Magnetic inductance (L_M), and the kinetic inductance is calculated equal to the kinetic energy stored in each conducting channel of the CNT. An effective kinetic inductance due to four parallel conducting channels in a CNT is equal to $L_K/4$. Here, the magnetic inductance is not taken into account due to large value of kinetic inductance. The equation of inductance given by (Rossi, Cazeaux, Metra, & Lombardi, 2007; Kumar, Majumder, Kaushik, Dasgupta, & Manhas, 2012)

$$L_K = \frac{h}{2e^2 v_F} \tag{7}$$

L_K (per unit length) for a CNT equals to 16 nH/um (Srivastava, Joshi, & Banerjee, 2005).

6.2. BUNDLED SWCNT

SWCNT bundle's geometrical structure with height *h* and width *w* is depicting in Figure 9, it consists of a number of SWCNTs of diameter *d* and spacing between the bundles' sp. The center to center distance of CNTs D=1.34 (Javey & Kong, 2009; Pandya, Majumder, Kaushik, & Manhas, 2012).The number of SWNTs in the bundle calculated as

$$n_W = \left| \frac{w - d}{sp} \right| + 1 \tag{8}$$

$$n_H = \left|\frac{h-d}{sp}\right| + 1 \tag{9}$$

$$n_{CNT} = n_W n_H - \left(\frac{n_H}{2}\right) \text{for } n_H \text{ is even number}$$

$$n_{CNT} = n_W n_H - \left(\frac{n_H - 1}{2}\right) \text{for } n_H \text{ is even number}$$

nH and *nW* are represented as the number of rows and columns in the bundle, *nCNT* is the total number of SWCNTs in the bundle. According to the geometry, the equivalent *RLC* electrical *circuit* model is shown in Figure 10.

6.2.1. CNT Bundle Resistance

The sum of the resistance of individual SWCNTs in bundle gives rise to the resistance of bundled SWCNTs. SWCNT has two types of resistances:

(1) Quantum resistance (*RQ*), equal to 6.45 (approx.) and (2) contact resistance, *RC* = 100 kΩ. The total resistance of SWCNT is given by (Kumar, Majumder, Kaushik, Dasgupta, & Manhas, 2012; Majumder, Kaushik, & Manhas, 2011; Pandya, Majumder, Kaushik, & Manhas, 2012),

$$R_{CNT} = R_C + R_Q\left(1 + {}^{l_{CNT}}\!/_{l_{mfp}}\right) \tag{10}$$

where l_{mfp} is represents the length of a mean free path and l_{CNT} is length of interconnect. It is presumed that all SWCNTs are placed parallel in the bundle. Therefore, the effective resistance of bundled SWCNT is described as the ratio of total bundle resistance to total numbers of SWCNTs present in bundle, expression given by,

$$R_{bundle} = {}^{R_{CNT}}\!/_{n_{CNT}} \tag{11}$$

$$R_{bundle} = (R_C + R_Q) / n_{CNT} + {}^{R_Q l_{CNT}}\!/_{n_{CNT} l_{mfp}} \tag{12}$$

nCNT is labeled as the total number of SWCNTs in bundle.

Figure 9. The geometrical structure of a bundled SWCNTs

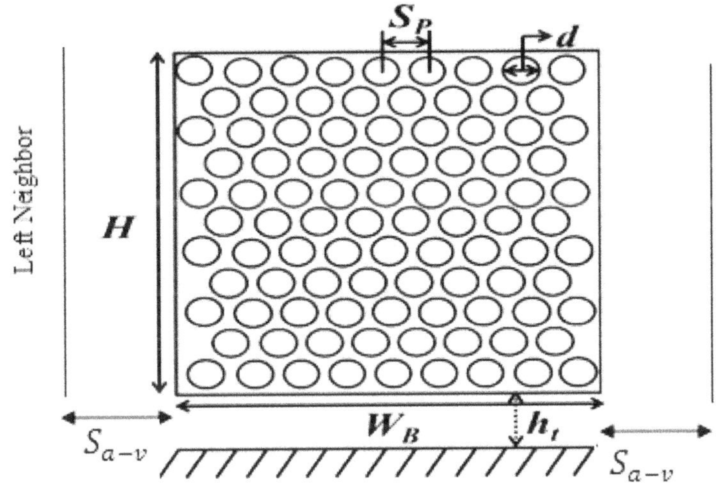

Figure 10. Equivalent RLC circuit model for two- bundled SWNT interconnect

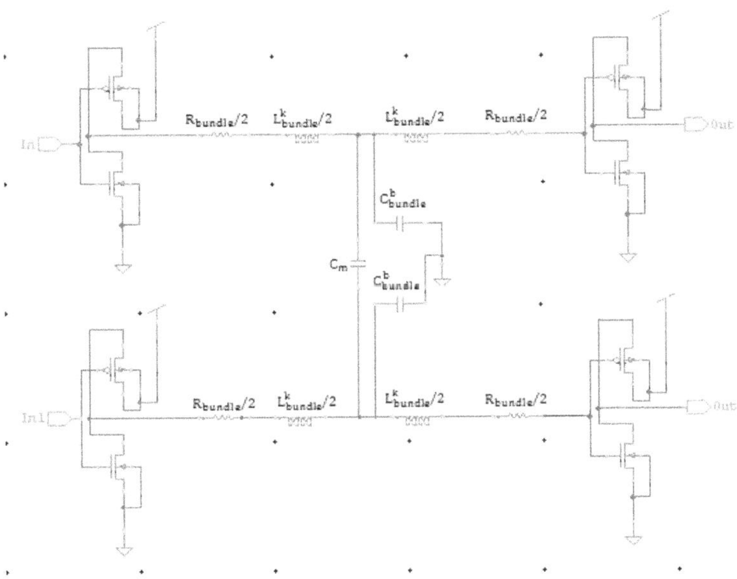

6.2.2. SWCNT Bundle Inductance

The inductance of bundle is given by (Kumar, Majumder, Kaushik, Dasgupta, & Manhas, 2012; Majumder, Kaushik, & Manhas, 2011)

$$\mathrm{L}_{\mathrm{Bundle}} = L_M^{Bundle} + L_K^{Bundle} \tag{13}$$

Where is $\mathrm{L}_{\mathrm{Bundle}}$ the Bundle inductance and L_M^{Bundle} , L_K^{Bundle} are magnetic, the kinetic inductance of bundle and given as (Srivastava, Joshi, & Banerjee, 2005)

$$L_M^{Bundle} = \frac{L_M}{n_{CNT}} \tag{14}$$

$$L_K^{Bundle} = \frac{L_K}{n_{CNT}} \tag{15}$$

The total inductance L_M^{Bundle} is given *as* the ratio of kinetic inductance to the total number of SWCNTs in a bundle. Magnetic inductance is neglected due to the large value of kinetic inductance.

6.2.3. SWCNT Bundle Capacitance

The coupling capacitance C_m between aggressor and victim is given below. Where, S_{a-v} is the spacing between the aggressor and victim line of bundled SWCNTs.

$$C_m = \frac{\pi \in}{\cosh^{-1}\left(\frac{S_{a-v}}{d}\right)} \tag{16}$$

The capacitance of bundle is given as (Kumar, Majumder, Kaushik, Dasgupta, & Manhas, 2012; Majumder, Kaushik, & Manhas, 2011)

$$\frac{1}{Cb_{Bundle}} = \frac{1}{C_Q^{Bundle}} + \frac{1}{C_{ES}^{Bundle}} \tag{17}$$

Where C_{Bundle} is bundle capacitance. C_Q^{Bundle}, C_E^{Bundle} are the quantum and electrostatic capacitance of bundle and given as

$$C_Q^{Bundle} = n_{CNT}\,(4.C_Q) \tag{18}$$

$$C_{ES}^{Bundle} = n_{CNT}\,(C_{ES}) \tag{19}$$

7. RESULTS AND DISCUSSIONS

The study of interconnects can be done by modeling its equivalent electrical circuit (the value of RLC is known from this circuit) and the crosstalk delay can be calculated (Singh Duksh, Kaushik, Sarkar, & Singh, 2013; Pandya, Majumder, Kaushik, & Manhas, 2012). The calculation of resistance, capacitance, and inductance of SWCNTs interconnects are described in following sections.

7.1. Resistance

The value of resistance has been calculated in kΩ for SWCNTs interconnects using equations (1-3) and (9-11). The values of resistance for bundled SWCNTs interconnect for different number of CNTs in a bundle for different length varying from 100um to 1000um at 32nm technology nodes. The values presented in the table given below.

From Table 1, it is observed that as the length of interconnecting increases the resistance increases. But resistance decreases with increase in number of SWCNTs in a bundle.

Table 1. The resistance of bundled SWCNTs for 127, 630 and 1128 tubes in a bundle

Resistance in kΩ			
Length (μm)	**127 SWCNTs in a Bundle**	**630 SWCNTs in a Bundle**	**1128 SWCNTs in a Bundle**
100	0.254	0.054	0.057
500	1.66	0.336	0.375
1000	2.54	0.51	0.573

7.2. Capacitance

The capacitances are of two types as studied in the chapter literature survey. The electrostatic capacitance and quantum capacitance comes in series combined get resultant capacitance by using equations (4-6) and (16-18).The value of capacitance between the interconnects is called coupling capacitance as C_{cs} depends upon the spacing between the bundled SWCNTs calculated by using equation (5).The values of both capacitance presented in table 2 below for 32nm technology.

As a result presented in Table 2 capacitance increases with increase in length of interconnects whereas, capacitance decreases with increasing the spacing between the SWCNTs bundles as shown in Table 3.

7.3 Inductance

Inductance has been calculated in nH for different length for SWCNT interconnects using equations (7) and (13-15). The values are listed below in Table 4 for different lengths and for different number of SWCNTs in a bundle.

Table 2. The capacitance C_{Bundle} of bundled SWCNTs for 127, 630 and 1128 tubes in a bundle

	Capacitance C_{Bundle} in fF		
Length (µm)	**127 SWCNTs in a Bundle**	**630 SWCNTs in a Bundle**	**1128 Swcnts In A Bundle**
100	4.995	4.999	4.99
500	24.975	24.99	24.99
1000	49.950	49.99	49.99

Table 3. Capacitance $\mathbf{C}_{m}$ of bundled SWCNTs for 127, 630 and 1128 tubes in a bundle

Capacitance C_{Bundle} in pF	
Spacing (nm)	**The Capacitance Between the Interconnects Lines**
2	21.193
3	15.84
4	13.67
5	12.17

Table 4. Inductance $L_{kBundle}$ of bundled SWCNTs for 127, 630 and 1128 tubes in a bundle

Inductance in nH			
Length (μm)	**127 SWCNTs in a Bundle**	**630 Swcnts In A Bundle**	**1128 SWCNTs in a Bundle**
100	1.57	0.317	0.1773
500	7.87	1.587	0.886
1000	15.748	3.17	1.773

As a result presented in Table 4, the inductance increases as increases in length of interconnects but decreases with increase in number of SWCNTs in a bundle.

8. SIMULATION PARAMETERS

Schematic of bundled SWCNTs have been drawn as L-type and T-type RLC network as two parallel line bus architecture; one is taken as aggressor and other as victim using S_Edit in last chapter. In this chapter, simulation is done using T_Spice. Parameters considered for simulation given below in Table 5. All the simulations performed at 32nm technology with width of NMOS =630nm and PMOS=1280. And voltage used is 1V. The results have been analyzed for different interconnects length, transition time and spacing between the lines.

Table 5. List of Parameters used for simulation

Diameter of CNT	1nm
Spacing between the SWCNTs bundles	2-5nm
Spacing between the CNT's	1.34nm
Frequency	1GHz,2GHz,5GHz, 10GHz
Input voltage	1V
Technology used	32nm
Width of NMOS,PMOS	640,1280
Number of SWCNTs in a bundle	127, 630,1128

Figure 11. Output waveform of equivalent circuit of SWCNTs as L-type RLC network

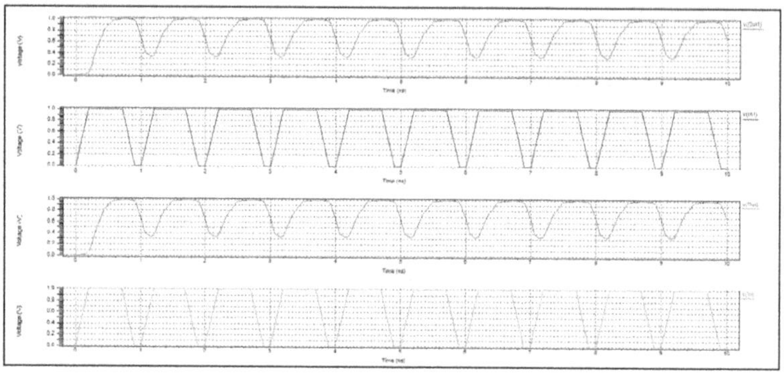

In Figure 11, the waveforms represent input and output of aggressor and victim lines:

- V(in) pink color waveform is input voltage for aggressor line.
- V(out) aqua color waveform is output voltage for aggressor line.
- V(in1) yellow color waveform is input voltage for victim line.
- V(out1) green output voltage for the victim line.

8. ANALYSIS OF CROSSTALK DELAY BETWEEN SWCNTS BUNDLES

8.1 Different Interconnects Length

In this section, the crosstalk delay will be analyzed between SWCNTs bundles with 127 CNTs in a bundle for different interconnects length with transition time 1ns and spacing between the

lines 2nm using 32 nm technology.

As simulation done at 32nm technology for different interconnects length by taking 127 SWCNTs in a bundle. From the simulation results, it has been analyzed that with increase in interconnects length the crosstalk delay increases with 30% from length 100nm to 1000 nm as shown in Figure 12.

Figure 12. Crosstalk delays for different interconnect length at 1ns transition time and 2 nm spacing between two lines (aggressor and victim)

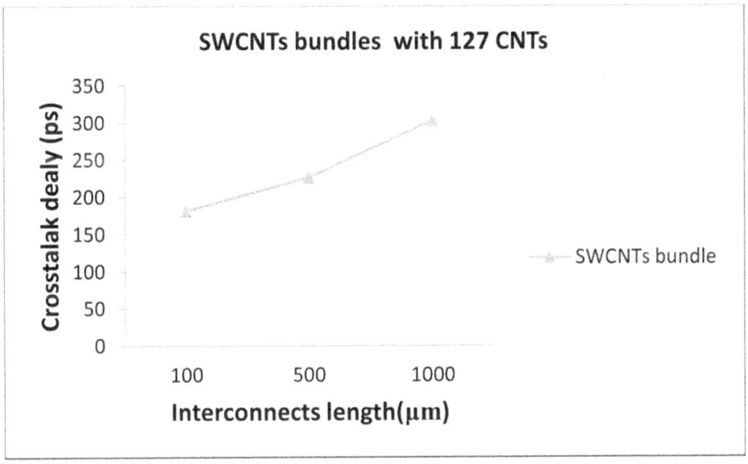

8.2 Different Transition Time

In this section, the crosstalk delay will be analyzed between SWCNTs bundles with 127 CNTs in a bundle for different transition time at interconnect length L=500µm and spacing between the lines 2nm using 32 nm technology.

As the simulation is done at 32nm technology for different transition time while keeping interconnect length=500µm and spacing between the SWCNTs bundles is 2nm by taking 127 SWCNTs in a bundle. From the simulation result, it can be seen through Figure 13 that crosstalk delay increases with increase in transition time. It means as we increase the frequency crosstalk delay will be less. Then Chip will be faster.

8.3 Different Spacing Between the Lines (Aggressor and Victim)

In this section, the crosstalk delay will be analyzed between SWCNTs bundles with 127 CNTs in a bundle for different spacing between the SWCNTs bundles by keeping interconnects length L=500µm and 1ns transition time.

As the simulation is done at 32nm technology for different spacing between the SWNTS bundles keeping interconnect L=500 and transition time 1ns. From the simulation result, crosstalk delay has been analyzed for different spacing between the SWCNTs parallel lines i.e. aggressor and victim. From Results represented in Figure 14, crosstalk delay decreases as increases the spacing between the two parallel lines of SWCNTs bundles.

Figure 13. Crosstalk delay for different transition time at 500μm interconnects length and 2 nm spacing between two lines (aggressor and victim)

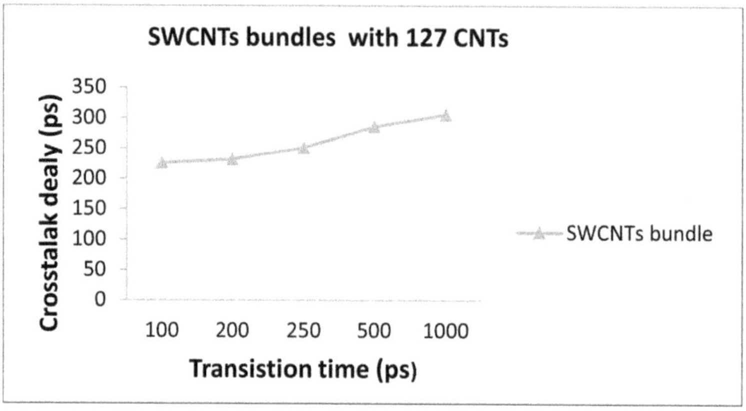

Figure 14. Crosstalk delay for different Spacing at 500μm interconnects length and 1ns transition time between two lines (aggressor and victim)

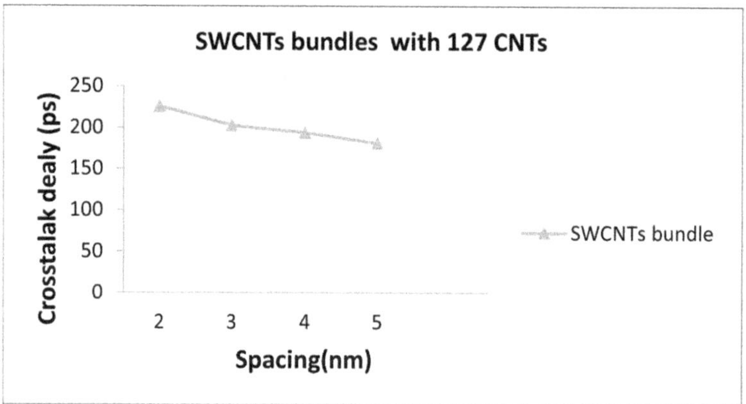

8.4 Number of SWCNTS in a Bundle

As described in previous sections, analysis of crosstalk delay is done by changing Length of interconnects, transition time and spacing between the two parallel lines of SWCNTs taking one as the aggressor and other is victim. it has been analyzed from simulation results that crosstalk delay increases as increases the length of interconnects, crosstalk delay also increases as increases transition time and crosstalk delay decreases as increases the spacing between the two parallel lines of SWCNTs. These effects were analyzed by taking 127 carbon nanotubes (CNTs) in a bundle for 32nm technology node. In this section, Crosstalk delay will be calculated for different number of carbon nanotubes (CNTs) in SWCNTs Bundle. The simulation will be done with 630 and 1128 carbon nanotubes (CNTs) in SWCNTs bundle for 32nm technology node.

8.5 Different Interconnects Length

In this section, the simulation results obtained from the T_Spice Tool. The number of CNTs is increased got from the calculation by increasing the width of bundle of SWCNTs using equations (4.8) and (4.9).The Bundles contains 630 and 1128 SWCNTs. Simulation results are analyzed for different interconnects length by keeping transition time 1ns and Spacing between the parallel lines 2nm fixed (i.e. aggressor and victim lines) will be analyzed. The values calculated from simulations results of crosstalk delay are given in Table 6 for different number of SWCNTs in a bundle. From Figure 15, it is cleared that as increases the number of CNTs in bundle crosstalk delay reduced by large value. It is much cleared from Table 6 and Figure 15, crosstalk delay is not much changed at low interconnects length.

But, it largely affected at long interconnect lengths. The average crosstalk delay reduced by 30% at 1000nm interconnects length as increases SWCNTs from 127 to 1128 in a bundle. Therefore, for global interconnects always used SWCNTs bundle with a large number of CNTs in the Bundle. As technology features size reduces interconnects plays very important role in VLSI Design.

8.6 Different Transition Time

Crosstalk delay increases as transition time increases. It was concluded that for high-frequency crosstalk delay is less at 32 nm deep submicron technology.

The values of crosstalk delay calculated given in table 7 at different transition time for 127, 630 and 1128 SWCNTs in a bundle. From the results, analyzed the crosstalk delay for 127,630 and 1128 carbon nanotubes (CNTs) with changing the transition time.

Table 6. Crosstalk delay at different interconnects length for 127, 630 and 1128 SWCNTs in a bundle

Crosstalk Delay (ps)			
Length (μm)	**127 SWCNTs in a Bundle**	**630 SWCNTs in a Bundle**	**1128 SWCNTs in a Bundle**
100	181.95	181.06	181.15
500	225.79	194.36	191.41
1000	302.82	216.3	210.45

Figure 15. Crosstalk delay at different interconnect length for 127,630, 1128 SWCNTs in bundles at 1ns transition time and 2 nm spacing between two lines (aggressor and victim)

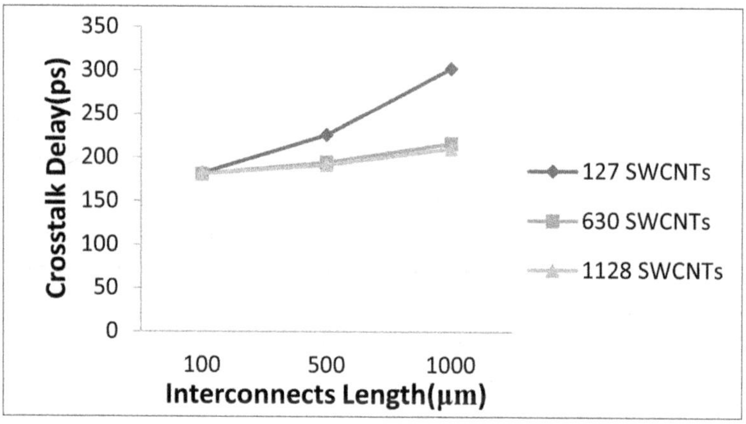

Table 7. Crosstalk delay at different Transition Time for 127,630, 1128 SWCNTs in a bundle

Crosstalk Delay (ps)			
Transition time (ps)	**127 SWCNTs in a Bundle**	**630 SWCNTs in a Bundle**	**1128 SWCNTs in a Bundle**
100	225.77	194.36	191.41
250	225.79	202.78	197.65
300	250.56	220.46	209.66
500	285.83	254.46	222.83
1000	305.51	260.53	255.45

By taking values of RLC according to the number of tubes in SWCNTs bundle. Simulation has been done by using T-Spice. Simulation is done by keeping interconnects Length L=500μm and spacing between the SWCNTs Bundles 2nm fixed using 32nm Technology node. From the simulation result, it has been concluded that crosstalk delay increases with increase in transition time and crosstalk delay decrease by 17% as increase in number of SWCNTs in a Bundle as shown in Fig 16 for 630 an 1128 CNTs in Bundles.

Figure 16. Crosstalk delay at different Transition Time for 127,630, 1128 SWCNTs in a bundle interconnects Length L=500μm and 2 nm spacing between two lines (aggressor and victim)

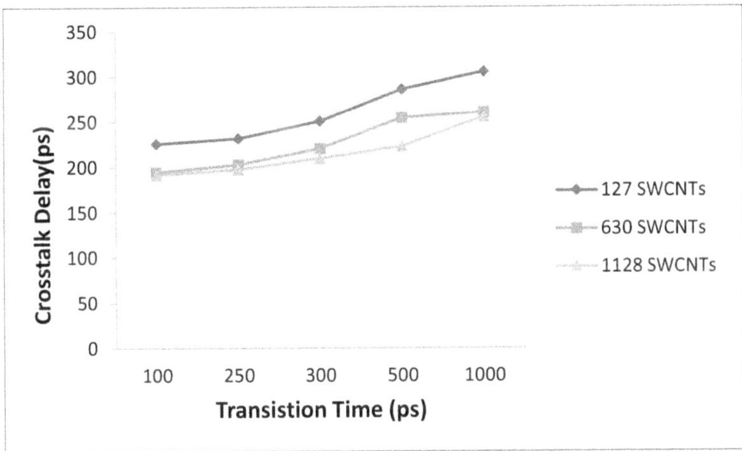

8.7 Different Spacing Between the Bundles

It was seen that with increases in spacing between the SWCNTs bundles crosstalk delay decreases. That simulation is done with 127 CNTs in the SWCNTs Bundle at 32 nm Technology node. The crosstalk delay calculated by simulation at different spacing between the lines for 127, 630 and 1128 SWCNTs in a bundle is given in Table 8.

From crosstalk delay values, here will analyze Crosstalk Delay with 630 and 1128 CNTs in SWCNTs in Bundles. Then will be compared among them. Graphical representation of simulation results of 127,630 and 1128 CNTs in SWCNTs bundles shown in Figure 17 below.

Table 8. Crosstalk delay at the different spacing between two lines (aggressor and victim) for 127,630, 1128 SWCNTs a in Bundle

	Crosstalk Delay (ps)		
Spacing (μm)	**127 Swcnts In A Bundle**	**630 SWCNTs in a Bundle**	**1128 SWCNTs in a Bundle**
2	225.03	194.25	190.86
3	224.45	194.05	190.08
4	223.91	193.85	189.71
5	222.86	193.5	188.86

Figure 17. Crosstalk delay at different spacing between two lines (aggressor and victim) for 127,630, 1128 SWCNTs in Bundles A interconnects Length L=500μm and 1ns transition time

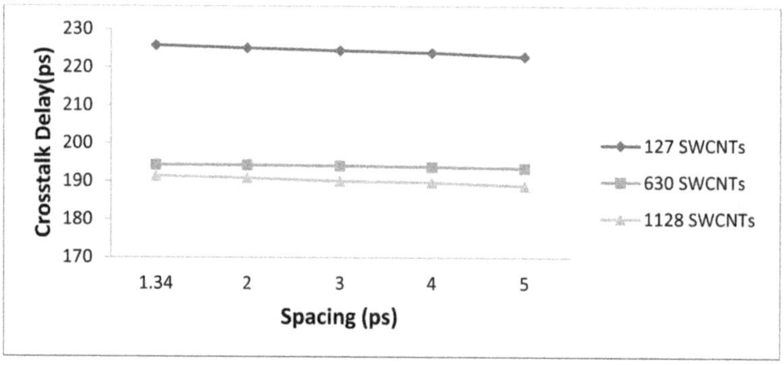

The Crosstalk Delay increases for both are containing 630 and 1128 CNTs in SWCNTs Bundles. It was also concluded that as the increase from 127 to 1128 CNTs in SWCNTs Bundles the crosstalk delay decreases by 15% for different spacing between the lines. So, for global interconnects bundle with large number of CNTs recommended for high frequency.

9. CONCLUSION

This chapter focused on the analysis of crosstalk delay of bundled SWCNT. Bundled SWCNT is presented with RLC equivalent model in parallel two-line bus architecture (aggressor and victim) using driver-interconnect-load system with CMOS driver at 32 nm technology node. From the simulation results, it has observed that crosstalk delay increases with increase in interconnect length and transition time. Whereas, it decreases with increase in spacing between the lines (aggressor and victim).

To reduce the crosstalk delay at global-level interconnect, the number of SWCNTs in a bundle has been increased. Thereby, values of parasitics i.e. resistance, capacitance, inductance also get reduced. Then, from simulation results, it is concluded that crosstalk delay reduced by 30% with increasing SWCNT bundle from 127 to 1128 at interconnects length, by 17% at different transition time and by 15% at different spacing between the lines (aggressor and victim) as increases number of SWCNTs from in a bundle.

Therefore, for global level interconnects, bundled SWCNTs are preferred for future VLSI technology.

10. FUTURE SCOPE

Carbon nanotubes (CNT) has emerged as most promising and attractive interconnect material for future deep sub-micron technology because of its phenomenal physical, mechanical and electrical properties i.e. high current carrying capacity, long ballistic transport, thermal stability, and tensile strength.

Crosstalk effects extensively affect the performance of digital design. So, further analysis of crosstalk delay for multi-wall carbon nanotube (MWCNTs) can be done. Then, on comparing the results of bundled SWCNTs and MWCNTs will be able to decide which one is best for global VLSI interconnects.

REFERENCES

Banerjee, K., & Srivastava, N. (2006, July). Are carbon nanotubes the future of VLSI interconnections? In *2006 43rd ACM/IEEE Design Automation Conference* (pp. 809-814). IEEE. 10.1145/1146909.1147116

Bhatnagar, S., Merkley, A., Berdine, R., Li, Y., & Roy, S. (2018, December). Variability-Aware Performance Assessment of Multi-Walled Carbon Nanotube Interconnects using a Predictor-Corrector Polynomial Chaos Scheme. In *2018 IEEE Electrical Design of Advanced Packaging and Systems Symposium (EDAPS)* (pp. 1-3). IEEE. 10.1109/EDAPS.2018.8680890

Das, D., & Rahaman, H. (2012, July). Unified model for analyzing timing delay and crosstalk effects in carbon nanotube interconnects. In *2012 4th Asia Symposium on Quality Electronic Design (ASQED)* (pp. 100-109). IEEE. 10.1109/ACQED.2012.6320484

Javey, A., & Kong, J. (Eds.). (2009). *Carbon nanotube electronics*. Springer Science & Business Media.

Kaushik, B. K., Sarkar, S., Agarwal, R. P., & Joshi, R. C. (2010). An analytical approach to dynamic crosstalk in coupled interconnects. *Microelectronics Journal*, *41*(2-3), 85–92. doi:10.1016/j.mejo.2009.12.011

Kumar, J., Majumder, M. K., Kaushik, B. K., Dasgupta, S., & Manhas, S. K. (2012, December). Novel modeling approach for multi-walled CNT bundle in global VLSI interconnects. In *2012 International Conference on Communications, Devices and Intelligent Systems (CODIS)* (pp. 476-479). IEEE. 10.1109/CODIS.2012.6422242

Kumar, M. G., Chandel, R., & Agrawal, Y. (2017). An efficient crosstalk model for coupled multiwalled carbon nanotube interconnects. *IEEE Transactions on Electromagnetic Compatibility*, *60*(2), 487–496. doi:10.1109/TEMC.2017.2719052

Loumi, A., Ghammaz, A., & Belahrach, H. (2019, April). Crosstalk and delay analysis in Multi-Walled Carbon Nanotube and Single-Walled Carbon Nanotube Interconnects. In *2019 International Conference on Wireless Technologies, Embedded and Intelligent Systems (WITS)* (pp. 1-4). IEEE. 10.1109/WITS.2019.8723814

Majumder, M. K., Das, P. K., Kaushik, B. K., & Manhas, S. K. (2012, December). Optimized delay and power performances for multi-walled CNT in global VLSI interconnects. In *2012 5th International Conference on Computers and Devices for Communication (CODEC)* (pp. 1-4). IEEE. 10.1109/CODEC.2012.6509358

Majumder, M. K., Kaushik, B. K., & Manhas, S. K. (2011, September). Comparison of propagation delay characteristics for single-walled CNT bundle and multiwalled CNT in global VLSI interconnects. In 2011 IEEE Recent Advances in Intelligent Computational Systems (pp. 911-916). IEEE.

Majumder, M. K., Kaushik, B. K., & Manhas, S. K. (2014). Analysis of delay and dynamic crosstalk in bundled carbon nanotube interconnects. *IEEE Transactions on Electromagnetic Compatibility*, *56*(6), 1666–1673. doi:10.1109/TEMC.2014.2318017

Majumder, M. K., Kaushik, B. K., & Manhas, S. K. (2014). Analysis of delay and dynamic crosstalk in bundled carbon nanotube interconnects. *IEEE Transactions on Electromagnetic Compatibility*, *56*(6), 1666–1673. doi:10.1109/TEMC.2014.2318017

Majumder, M. K., Kaushik, B. K., Manhas, S. K., & Das, P. K. (2012, December). Dynamic crosstalk effect in CNT bus architecture. In *2012 International Conference on Emerging Electronics* (pp. 1-4). IEEE. 10.1109/ICEmElec.2012.6636252

Majumder, M. K., Pandya, N. D., Kaushik, B. K., & Manhas, S. K. (2012, March). Analysis of crosstalk delay and area for MWNT and bundled SWNT in global VLSI interconnects. In *Thirteenth International Symposium on Quality Electronic Design (ISQED)* (pp. 291-297). IEEE. 10.1109/ISQED.2012.6187508

Majumder, M. K., Pandya, N. D., Kaushik, B. K., & Manhas, S. K. (2012). Dynamic crosstalk effect in mixed CNT bundle interconnects. *Electronics Letters*, *48*(7), 384–385. doi:10.1049/el.2012.0536

Majumder, M. K., Pandya, N. D., Kaushik, B. K., & Manhas, S. K. (2012). Analysis of MWCNT and bundled SWCNT interconnects: Impact on crosstalk and area. *IEEE Electron Device Letters*, *33*(8), 1180–1182. doi:10.1109/LED.2012.2200872

Majumder, M. K., Pandya, N. D., Kaushik, B. K., & Manhas, S. K. (2012). Comparison of crosstalk delay between single and bundled SWNT for global VLSI interconnects. In *1st International Conference on Recent Advances in Information Technology (RAIT)*, (pp. 564-569), IEEE. 10.1109/RAIT.2012.6194590

Mittal, J., & Lin, K. L. (2017). Carbon nanotube-based interconnections. *Journal of Materials Science, 52*(2), 643–662. doi:10.100710853-016-0416-4

Naeemi, A., & Meindl, J. D. (2006). Design and performance modeling for single-walled carbon nanotubes as local, semiglobal, and global interconnects in gigascale integrated systems. *IEEE Transactions on Electron Devices, 54*(1), 26–37. doi:10.1109/TED.2006.887210

Pandya, N. D., Majumder, M. K., Kaushik, B. K., & Manhas, S. K. (2012, May). Performance Comparision of Mixed CNT Bundle in Global VLSI Interconnect. In *2012 International Conference on Communication Systems and Network Technologies* (pp. 790-793). IEEE. 10.1109/CSNT.2012.171

Pu, S. N., Yin, W. Y., Mao, J. F., & Liu, Q. H. (2009). Crosstalk prediction of single- and double-walled carbon-nanotube (SWCNT/DWCNT) bundle interconnects. *IEEE Transactions on Electron Devices, 56*(4), 560–568. doi:10.1109/TED.2009.2014429

Rossi, D., Cazeaux, J. M., Metra, C., & Lombardi, F. (2007). Modeling crosstalk effects in CNT bus architectures. *IEEE Transactions on Nanotechnology, 6*(2), 133–145. doi:10.1109/TNANO.2007.891814

Singh Duksh, Y., Kaushik, B. K., Sarkar, S., & Singh, R. (2013). Analysis of propagation delay and power with variation in driver size and number of shells in multi walled carbon nanotube interconnects. *Journal of Engineering, Design and Technology, 11*(1), 19-33.

Srivastava, N., & Banerjee, K. (2005, May). Performance analysis of carbon nanotube interconnects for VLSI applications. In *Proceedings of the 2005 IEEE/ACM International conference on Computer-aided design* (pp. 383-390). IEEE Computer Society. 10.1109/ICCAD.2005.1560098

Srivastava, N., Joshi, R. V., & Banerjee, K. (2005, December). Carbon nanotube interconnects: implications for performance, power dissipation and thermal management. In IEEE InternationalElectron Devices Meeting, 2005. IEDM Technical Digest. (pp. 249-252). IEEE. doi:10.1109/IEDM.2005.1609320

Uhlig, B., Dhavamani, A., Nagy, N., Lilienthal, K., Liske, R., Ramos, R., & Georgiev, V. (2018, June). Challenges and Progress on Carbon Nanotube Integration for BEOL Interconnects. In *2018 IEEE International Interconnect Technology Conference (IITC)* (pp. 16-18). IEEE. 10.1109/IITC.2018.8454842

Uhlig, B., Liang, J., Lee, J., Ramos, R., Dhavamani, A., Nagy, N., . . . Asenov, A. (2018, March). Progress on carbon nanotube BEOL interconnects. In 2018 Design, Automation & Test in Europe Conference & Exhibition (DATE) (pp. 937-942). IEEE. doi:10.23919/DATE.2018.8342144

Wei, B. Q., Vajtai, R., & Ajayan, P. M. (2001). Reliability and current carrying capacity of carbon nanotubes. *Applied Physics Letters*, *79*(8), 1172–1174. doi:10.1063/1.1396632

Chapter 9
Performance Analysis of Mixed-Wall CNT Interconnects Using Colliding Bodies Optimization Technique

Girish Kumar Mekala
Vidya Jyothi Institute of Technology, India

Yash Agrawal
Dhirubhai Ambani Institute of Information and Communication Technology, Gandhinagar, India

Rajeevan Chandel
National Institute of Technology, Hamirpur, India

Ashwani Kumar
National Institute of Technology, Hamirpur, India

ABSTRACT

In recent years, carbon nanotube (CNT) interconnects have emerged as a potential alternative to copper interconnects due to their several magnificent properties. Due to fabrication issues, realization of densely packed CNTs with uniform diameters in a bundle structure is difficult to achieve. Consequently, it is advantageous to obtain a combination of CNTs with non-uniform diameters in the bundle, thereby leading to a densely packed mixed-wall CNT bundle (MWCB). In a MWCB structure, tube density plays a major role to determine the parasitic elements associated with the interconnects. For this, prospectively, colliding bodies optimization (CBO) technique has been incorporated. It is inferred from the study that the overall crosstalk noise, delay, and power dissipation of MWCB interconnect with higher tube density (i.e., obtained using CBO technique) are lesser than other CNT structures. Henceforth, it is determined from the proposed work that prospective CBO technique for advanced MWCB structure is highly efficient and effective for on-chip interconnects in IC designs.

DOI: 10.4018/978-1-7998-1393-4.ch009

1. INTRODUCTION

In recent years, carbon nanotube (CNT) interconnects have emerged as a potential alternative to copper interconnects due to their significantly better electrical and thermal properties (Novoselov, 2004; Yu, 2000; Wei, 2001; Hone, 1999). CNTs can be broadly classified into single- and multi-walled nanotubes (SWCNTs and MWCNTs). The conductivity of individual SWCNTs is quite low due to high quantum resistance. Hence, to improve the conductivity of SWCNTs, bundle structures have been proposed by several researchers (Koo, 2007; Pu, 2009; Srivastava, 2009; Li, 2009). However, considering the wrapped direction of graphene layers, SWCNTs can be either metallic or semiconducting, whereas, MWCNTs always remain metallic due to their large diameters (Li, 2009).

Modeling and performance analyses of SWCNT and MWCNT interconnect to examine propagation delay, power dissipation, and crosstalk have been reported in literature (Giustiniani, 1986; Sarto, 2010; Kumar, 2015; Agrawal, 2016; Kumar, 2016). It is envisaged that fabrication of CNTs with uniform diameter is difficult to attain. Consequently, the CNT bundle comprising of CNTs with varying diameter configuration is considered (Zhu, 2006; Chenug, 2002). MWCBs possess relatively high conductance than those of the SWCNT bundle or a MWCNT (Haruehanroengra, 2007). In Wang (2007), have investigated that tube density is an important entity to determine the parasitics of on-chip interconnects. The work in Majumder (2014) shows that crosstalk induced delay can be significantly minimized by using high tube density. It is seen that the MWCB structure comprises of random size non uniform CNTs placed at different locations. This creates lot of free space within the MWCB structure (Yilmazoglu, 2011). Higher the free space, lower with the tube density. This decreases the performance of the system. Consequently, optimization of MWCB structure is essentially needed so that free space can be minimized and henceforth tube density can be improved. This can be attained by proper placing of CNT from larger to smaller diameter CNTs in MWCB structure.

In this research work, prospective colliding bodies optimization (CBO) scheme is incorporated to determine optimum tube density for MWCB on-chip interconnects. The performance analysis is performed at 22nm technology. The CBO algorithm has been widely explored by different researchers (Kaveh, 2014; Panda, 2014). This has been vividly adopted to solve different engineering problems. The optimization scheme and optimal tube density for MWCB structure have been very sparsely reported in literature yet. CBO technique is based on 1-dimensional collision between two adjacent objective bodies and it does not involve any tunable parameters (Kaveh, 2014). Along with using CBO algorithm, there is need to fit different CNTs within the rectangular shape of MWCB structure. The effective placement of CNTs in MWCB structure is performed using Huang's approach (Huang, 2004). This approach determines the procedure of optimally place circle with non-uniform diameter in to a rectangular space.

2. RANDOMIZED PLACEMENT OF CNTS IN MIXED WALL CNT BUNDLE

The schematic of MWCB comprising of randomized distribution of CNTs with varying diameter having height (*H*) and width (*W*) is illustrated in Figure 1. The MWCB is employed above a ground plane at height h_t. The arrangement of different CNTs are in 2D Euclidean space that has its origin at the center. The inter gap between adjacent CNTs is 0.34nm (Li, 2009). The MWCB interconnect comprises of both MWCNTs and SWCNTs. This is shown in Figure 1(a). Figure 1(b) illustrates the concentric shells of MWCNT interconnects. The MWCNT outermost shell diameter (d_p) can be determined as (Kumar, 2016):

$$d_p = d_1 + 2 \times \delta \times (p-1) \tag{1}$$

where, d_1 denotes the diameter of inner-most shell, p corresponds to total shells umber, δ represents Vander-Waal's gap.

The diameters of MWCNTs are defined as d_p, d_{p-1}, d_{p-2}, …, d_2 corresponding to respective p, p-1, p-2,…, 2 shells. The diameter of outermost shell of SWCNT is denoted by d_1. This comprises of a single shell. The placement of CNTs in MWCB structure is from larger diameter to smaller diameter. This is followed in random manner for obtaining higher number of conducting channels. To efficiently implement this, Huang's approach is utilized as depicted in Fig. 1 (c) (Huang, 2004; Majumder,

Figure 1. (a) Placement of MWCNTs/SWCNTs in MWCB interconnect (b) Cross sectional view of an MWCNT (c) Huang's approach representation

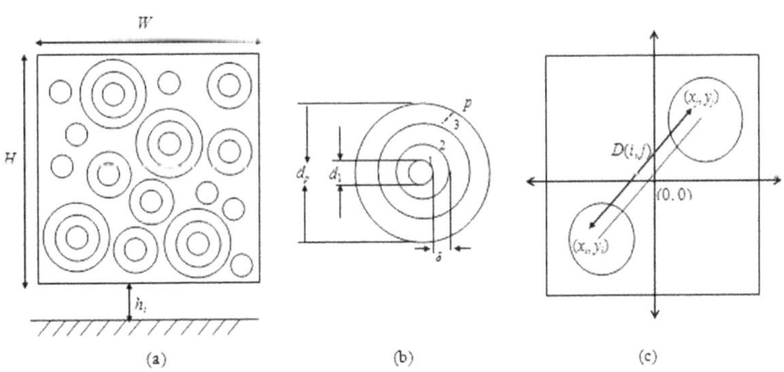

2015). The Huang's approach defined the set boundary conditions. These are defined in equation (2)-(5). Equation (6) is used to ensure that there is no overlap between two CNTs.

$$\frac{H}{2} - y_i - \frac{d_i}{2} \geq 0; \qquad i \in \{1,2,\cdots,p\} \tag{2}$$

$$\frac{H}{2} + y_i - \frac{d_i}{2} \geq 0; \qquad i \in \{1,2,\cdots,p\} \tag{3}$$

$$\frac{W}{2} + x_i - \frac{d_i}{2} \geq 0; \qquad i \in \{1,2,\cdots,p\} \tag{4}$$

$$\frac{W}{2} - x_i - \frac{d_i}{2} \geq 0; \qquad i \in \{1,2,\cdots,p\} \tag{5}$$

where d_1 and d_2 are diameter of two CNTs that are placed at coordinates (x_i, y_i) and (x_j, y_j) respectively. $D(i, j)$ defines the center-to-center distance between two CNTs in MWCB structure.

$$D(i,j) = \sqrt{(x_i - x_j)^2 + (y_i - y_j)^2} - \frac{d_i}{2} - \frac{d_j}{2} \geq 0; \qquad i \neq j \in \{1,2,\cdots,p\} \tag{6}$$

Figure 2. Multi transmission line model of MWCB interconnect

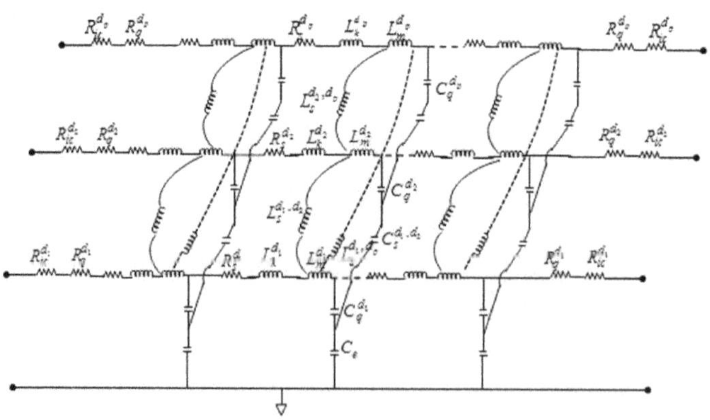

Figure 3. Representation of equivalent RLC model of MWCB interconnect

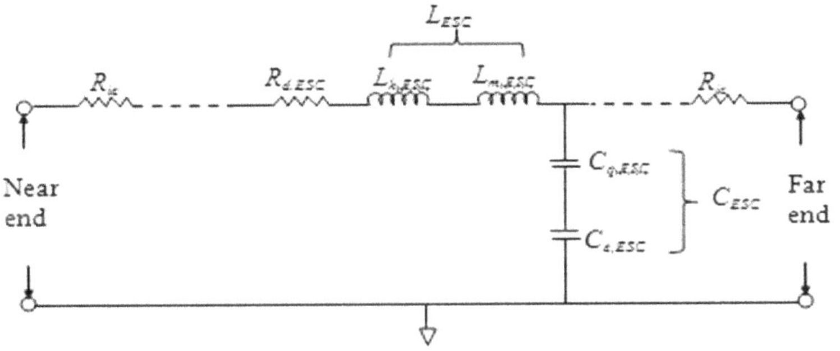

The multi transmission model of MWCB interconnects with is given in Figure 2. All SWCNTs and MWCNTs are assumed to be in parallel and subsequently, simplified into an equivalent *RLC* model. This is depicted in Fig. 3. The value of contact resistance (R_{ic}) varies with the fabrication process (Majumder, 2014; Srivastava, 2010). The CNT comprises of quantum resistance (R_q) and it is due to low energy levels in nano interconnects. The other resistive component of CNT is scattering resistance ($R_{d,ESC}$) is due to surface, edge and phonon scatterings (Novoselov, 2004; Yu, 2000; Wei, 2001; Hone, 1999).

$$R_{d,ESC} = \left[\left(\frac{1}{N_{d_1}} \left(\frac{R_q}{\sum_{s=1}^{p} N_{ch,s} \cdot l_{mfp,s}} \right) \right)^{-1} + \cdots\cdots + \left(\frac{1}{N_{d_p}} \left(\frac{R_q}{\sum_{s=1}^{p} N_{ch,s} \cdot l_{mfp,s}} \right) \right)^{-1} \right]^{-1} \tag{7}$$

where,

$$R_q = \frac{h}{2e^2} \tag{8}$$

$N_{d_1}, N_{d_2}, \ldots, N_{d_p}$ denote count of CNTs with diameter as $d_1, d_2, \ldots, d_p$ respectively. N_{ch} is the number of conducting channels [8]. l_{mfp}, h and e denote the mean free path, Planck's constant and electronic charge respectively.

The CNT comprises of inductive components. This includes kinetic ($L_{k,ESC}$) and magnetic ($L_{m,ESC}$) inductance. Both these inductive components are shown in Figure 3. The kinetic inductance is due to kinetic energy associated with the charge carriers and flow of current while magnetic inductance arises due to flow of charge carriers through CNT (Wang, 2007). The per unit length (p.u.l) of kinetic and magnetic inductance are given as (Majumder, 2015):

$$L_{k,ESC} = \left[\left(\frac{1}{N_{d_1}} \left(\frac{l_k}{\sum_{s=1}^{p} N_{ch,s}} \right) \right)^{-1} + \cdots + \left(\frac{1}{N_{d_p}} \left(\frac{l_k}{\sum_{s=1}^{p} N_{ch,s}} \right) \right)^{-1} \right]^{-1} \tag{9}$$

where,

$$l_k = \frac{h}{2e^2 v_F} \tag{10}$$

$$L_{m,ESC} = \frac{\mu_0}{2\pi} \cosh^{-1} \left(\frac{d_p + 2h_t}{d_p} \right) \tag{11}$$

where, the CNT Fermi velocity is v_F=8×10^5 m/s.

The MWCB consist of electrostatic ($C_{e,ESC}$) and quantum ($C_{q,ESC}$) capacitances. The electrostatic capacitance is due to creation of electric potential difference between ground place and MWCB. The quantum capacitance arises owing to quantum energy state of charge carrier. The per unit length (p.u.l.) of electrostatic and quantum capacitance are expressed as (Majumder, 2015):

$$C_{ESC} = \left(\frac{1}{C_{q,ESC}} + \frac{1}{C_{e,ESC}} \right)^{-1} \tag{12}$$

$$C_{q,ESC} = \frac{2e^2}{hv_F} \left[N_{d_1} \sum_{s=1}^{p} N_{ch,s} + \cdots + N_{d_p} \sum_{s=1}^{p} N_{ch,s} \right] \tag{13}$$

$$C_{e,ESC} = \frac{2\pi\varepsilon_0\varepsilon_r}{\cosh^{-1}\left[\left(d_p + 2h_t \right) / d_p \right]} \times N_x \tag{14}$$

where, ε_r and N_x are the relative permittivity of dielectric medium and the number of CNTs towards the ground (ITRS, 2012). In addition, each shell and each SWCNT/MWCNT in a MWCB structure experiences shell-shell coupling capacitance (C_s) and inter-CNT mutual capacitance (C_m) (Sarto, 2010; Majumder, 2014). The mutual inductance between the two interconnects is calculated in accordance to (Agrawal, 2016).

It is inferred from (7) to (14) that number of CNT along with conduction channels affect the parasitics of MWCB interconnect. The MWCB parasitics can be minimized by incorporating more CNTs with larger diameter. This enables in creation of more number of conduction channels. Tube density is one of the efficient parameter to determine packaging density and number of conduction channels. This is determined as:

$$\text{Tube density } (F) = \frac{\sum_{s=1}^{p} s \times N_{d_s}}{W \times H} \tag{15}$$

where, s denotes the shell number that is varying from 1 to p.

One of the main objective of the present work is to enhance the performance of nano MWCB interconnects. This is prospectively attained using CBO technique. This is discussed next in this paper.

3. COLLIDING BODIES OPTIMIZATION (CBO)

Kaveh et al. (2014) have introduced colliding bodies optimization (CBO) algorithm for solving complex optimization problems. CBO is multi-agent technique, based 1-dimensional collisions that take place between any two bodies. In this approach, each agent corresponds to colliding body (CB) that attributes certain velocity and mass. The positions of each of the CBs are successively updated depending on the collision laws as presented below.

3.1 Laws of Momentum and Energy

Laws of energy and momentum govern the collisions and the momentum exchange between any two bodies. When a collision happens in a system under absence of any external forces, net momentum of the bodies after and before the collision remains the same. Figure 4 shows the positions of bodies before and after collision. The same can be expressed as follows (Kaveh, 2014):

$$m_1 v_1 + m_2 v_2 = m_1 v'_1 + m_2 v'_2 \tag{16}$$

The total kinetic energy conservation of such a system is represented as follows:

$$\frac{1}{2} m_1 v_1^2 + \frac{1}{2} m_2 v_2^2 = \frac{1}{2} m_1 v'^2_1 + \frac{1}{2} m_2 v'^2_2 + Q \tag{17}$$

where, m_1 and m_2 denote the mass of the 1st and 2nd body respectively. v_1 and v_2 represent the 1st and 2nd body velocities before the collision. v'_1 and v'_2 denotes the 1st and 2nd body velocities after the collision. Q denotes the loss in kinetic energy due to collision.

The body velocities after the collision are given as:

$$v'_1 = \frac{(m_1 - \alpha m_2) v_1 + (m_2 + \alpha m_2) v_2}{m_1 + m_2} \tag{18}$$

$$v'_2 = \frac{(m_2 - \alpha m_1) v_2 + (m_1 + \alpha m_1) v_1}{m_1 + m_2} \tag{19}$$

where, α represents the coefficient of restitution of the colliding bodies and is defined as the ratio of relative velocities of the bodies after collision to those before the collision (Kaveh, 2014; Kaveh, 2014; Panda, 2014).

Collisions can be classified as elastic (if $\alpha=1$) and inelastic collisions (if $\alpha<1$).

The elastic collisions there is no loss in kinetic energy while in case of inelastic collisions, part of the kinetic energy gets transformed in other form of energy. The separative velocities are high in elastic collisions while it is low in inelastic collisions.

3.2 Eminent Algorithm Development for the Novel MWCB Structure

Step 1: Initialization and defining *H, W* and d_p parameters. The first step comprises of defining CNTs positions. The CNTs are also referred to as agents or colliding bodies. The positions of the agents are initially randomly defined. The randomized location of CNTs are iteratively tested multiple number times. This refers to population size. The complete set of CB position for carrying population is examined as:

Figure 4. Illustration of two moving bodies (a) before and (b) after collision

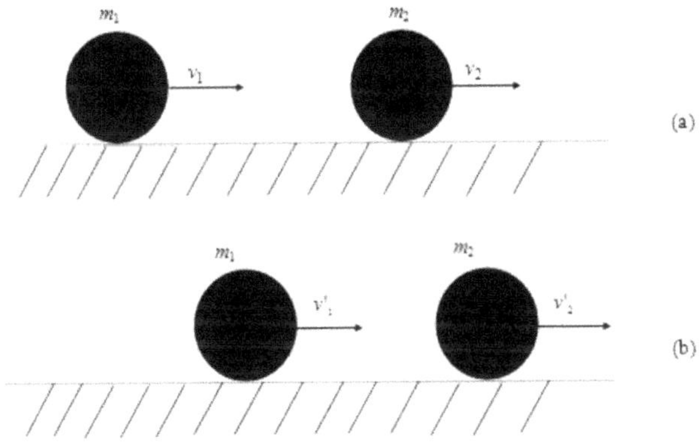

$$X=\begin{bmatrix}X_1\\X_2\\\vdots\\X_k\\\vdots\\X_{NP}\end{bmatrix}=\begin{bmatrix}(x_{1,1},y_{1,1}),(x_{1,2},y_{1,2}),\cdots,(x_{1,m},y_{1,m}),\cdots,(x_{1,NA},y_{1,NA})\\(x_{2,1},y_{2,1}),(x_{2,2},y_{2,2}),\cdots,(x_{2,m},y_{2,m}),\cdots,(x_{2,NA},y_{2,NA})\\\vdots\quad\vdots\quad\vdots\quad\vdots\\(x_{k,1},y_{k,1}),(x_{k,2},y_{k,2}),\cdots,(x_{k,m},y_{k,m}),\cdots,(x_{k,NA},y_{k,NA})\\\vdots\quad\vdots\quad\vdots\quad\vdots\\(x_{NP,1},y_{NP,1}),(x_{NP,2},y_{NP,2}),\cdots,(x_{NP,m},y_{NP,m}),\cdots,(x_{NP,NA},y_{NP,NA})\end{bmatrix} \tag{20}$$

where, k=1,2,…, NP and m=1,2,…, NA. NP and NA are number of populations and number of agents respectively.

Step 2: In the second step, CNTs are sorted and placed in accordance to them larger to smaller diameter.

Step 3: the third step comprises of checking the viability of the position at which CNTs are placed. This is performed similarly Huang's approach as per the expressions presented in (2) to (6).

Step 4: Further on, the process is repeated from largest diameter to smallest diameter.

Step 5: Once all CNTs are placed, the fitness value is computed for the entire MWCB structure using (15).

Step 6: Subsequently, for each CB, body mass is evaluated as

$$m_k = \frac{1/F(k)}{\sum_{j=1}^{NP} 1/F(j)} \tag{21}$$

where, $F(k)$ and $F(j)$ corresponds to fitness values of the kth and j^{th} CB respectively.

Step 7: Once mass values of all CBs are computed, these are equally segmented in two categories as stationary bodies and moving bodies as shown in Figure 5. The velocities of the stationary and moving bodies before the collision are obtained as

$$v_k = 0; \; k = 1, 2 \ldots, \frac{NP}{2} \tag{22}$$

$$v_k = X_{k-\frac{NP}{2}} - X_k; \; k = \frac{NP}{2} + 1, \ldots, NP \tag{23}$$

where, X_k is the position vector of the kth CB.

Step 8: It is assured that there is movement of moving bodies towards stationary bodies. As a result there is collision between the moving and stationary bodies. Consequently, the velocities of stationary and moving bodies changes after collision and are expressed as

$$v_k' = \frac{\left(m_{k+\frac{NP}{2}} + \alpha m_{k+\frac{NP}{2}}\right) v_{k+\frac{NP}{2}}}{m_k + m_{k+\frac{NP}{2}}} \quad k = 1, 2 \ldots, \frac{NP}{2} \tag{24}$$

$$v_k' = \frac{\left(m_k - \alpha m_{k-\frac{NP}{2}}\right) v_k}{m_k + m_{k-\frac{NP}{2}}} \quad k = \frac{NP}{2} + 1, \ldots, NP \tag{25}$$

where, α is calculated as below

$$\alpha = 1 - \frac{J}{J_{\max}} \tag{26}$$

Figure 5. Representation of stationary and moving colliding bodies

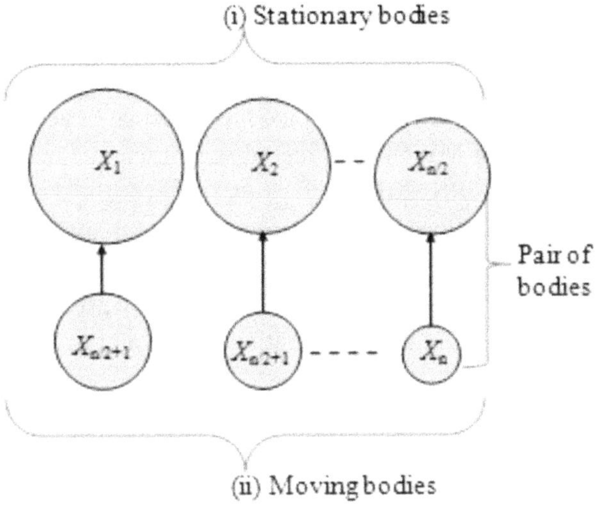

where, J denote the count for present iteration. J_{max} represents the maximum number of iterations respectively.

Step 9: Corresponding to obtained new velocities, the respective position of each agent is updated using (27) and (28).

$$X_k^{J+1} = X_k^J + rand(1, NA) * V_k^J \ ; \ k = 1, 2, \ldots, \frac{NP}{2} \tag{27}$$

$$X_k^{J+1} = X^J_{k-\frac{NP}{2}} + rand(1, NA) * V_k^J \ ; \ k = \frac{NP}{2} + 1, \ldots, NP \tag{28}$$

where, X_k^J and X_k^{J+1} denote the position vectors corresponding to the kth CB for J^{th} and $(J+1)^{th}$ iterations respectively.

Step 10: In case the CB hits its limits, then there are correspondingly set to boundary value.

Step 11: the process is executed continuously unless it reach the maximum iteration count. Once it is achieved the maximum tube density and corresponding count of CNTs are computed. The maximum iteration value is not attained, then process is repeated again from step 3.

4. SETUP AND SIMULATIONS

The set up for MWCB interconnect comprises of CNTs with varying shells. The count of shells is varying from 1 to 10. The innermost shell diameter is 1nm. Using (1), the outermost shell diameter is computed as 7.12nm. In accordance to ITRS, the thickness and width of MWCB considered as 75nm and 32nm respectively for 22nm technology node (ITRS, 2012).

The driver-interconnect-load model is used for present analysis where driver is CMOS inverter. The MWCB structure is considered as interconnect when load is a CMOS inverter which is approximated as capacitor of 2fF. The transition time of input signal is 20ps. The number of iterations and count of CBs considered are 100 and 50 respectively.

For the current analysis, 500 independent trials have been implemented using CBO method for random distribution of MWCNT (1.68nm-7.12nm) and SWCNTs (1nm) in a MWCB. Initially, 50 independent trials were performed. Further the trials were increased up to 1000. After repeated trials, 500 independent trails have been selected, as this satisfy the parameters of the analysis. These trials provide, the maximum, average and the minimum tube density values along with number of CNTs for each diameter are obtained. The maximum, average and minimum tube density CNT bundle structures are henceforth referred as MWCB-I, MWCB-II and MWCB-III respectively. These models are compared with the tube density of MWCB structure (MWCB-IV) reported in the literature (Majumder, 2015). The corresponding values are tabulated in Table 1. The value of tube density that are obtained for MWCB-I, MWCB-II and MWCB-III are $23.95\times10^{12}/cm^2$, $22.66\times10^{12}/cm^2$ and $21.45\times10^{12}/cm^2$ respectively. For assessing the robustness of presented CBO technique, two essential parameters viz. standard deviation and percentage of successful convergence rate have been reported in the literature (Rao, 1984). The maximum objective function value (i.e. $23.95\times10^{12}/cm^2$) of the problem close to average value (i.e. $22.66\times10^{12}/cm^2$) confirms the reliability and robustness of the CBO technique. The other effective method is percentage of successful convergence to determine robustness of CBO technique. The convergence rate is defined as the count of runs that successfully converge (i.e. maximum tube density) to the maximum count of trails. In the present case, it is seen that out of total 500 trials, 460 trials successfully converged. This corresponds to convergence rate of 92%. This high convergence rate reflects the robustness of CBO technique.

The number of different diameter CNTs facing the ground plane for different bundle configurations under consideration are given in Table 2. In Table 3, *RLC* values are evaluated as per (7)-(14).

Table 1. Count of CNTS in MWCB structure with different tub densities

Bundle Structure	Various diameter CNTs in the cross-sectional area (32nm×75nm) of MWCB structure										Tube density (×10¹²/cm²)
	s=1	s=2	s=3	s=4	s=5	s=6	s=7	s=8	s=9	s=10	
	d_1=1 nm	d_2=1.68 nm	d_3=2.36 nm	d_4=3.04 nm	d_5=3.72 nm	d_6=4.4 nm	d_7=5.08 nm	d_8=5.76 nm	d_9=6.44 nm	d_{10}=7.12 nm	
MWCB-I	53	22	17	6	3	7	3	4	3	24	23.95
MWCB-II	44	19	12	9	3	3	3	4	2	24	22.66
MWCB-III	33	21	5	6	5	7	5	3	1	22	21.45
MWCB-IV [24]	19	12	3	8	4	4	1	3	2	21	17.87

Table 2. Various diameter CNTs towards ground in MWCB structure

CNT Bundle	Various diameter CNTs towards the ground (N_x) having a diameter of									
	1nm	1.68nm	2.36nm	3.04nm	3.72nm	4.4nm	5.08nm	5.76nm	6.44nm	7.12nm
MWCB-I	2	1	1	0	1	0	0	1	0	4
MWCB-II	11	3	0	3	0	1	1	0	0	0
MWCB-III	8	0	4	0	2	2	0	0	0	1
MWCB-IV [24]	17	2	1	0	1	0	1	1	0	4

Table 3. RLC parameters for different MWCB structure

CNT Bundle	*RLC* Parameters					
	$R_{ic,ESC}$ (Ω)	$R_{s,ESC}$ (Ω/μm)	L_{ESC} (pH/μm)	M_{ESC} (pH/μm)	C_{ESC} (aF/μm)	$C_{c,ESC}$ (aF/μm)
MWCB-I	38.1	4.85	16.20	0.72	24.25	32.87
MWCB-II	39.3	5.25	17.96	0.86	43.18	48.86
MWCB-III	39.6	5.98	18.74	0.95	51.76	56.27
MWCB-IV [24]	41.5	6.76	22.35	1.12	68.53	71.29

As the reliability of MWCB interconnects depends upon power dissipation, propagation delay, and crosstalk. Fig. 6 shows n-coupled MWCB interconnect lines to evaluate crosstalk analysis. Crosstalk noise can be classified as functional and dynamic. In the former case, all inputs switch simultaneously except the input of victim line (which is quite). Whereas, in the latter case, the input of both aggressor as well victim line switch in either out of phase or in-phase with each other.

Figure 6. n-Coupled MWCB interconnect lines

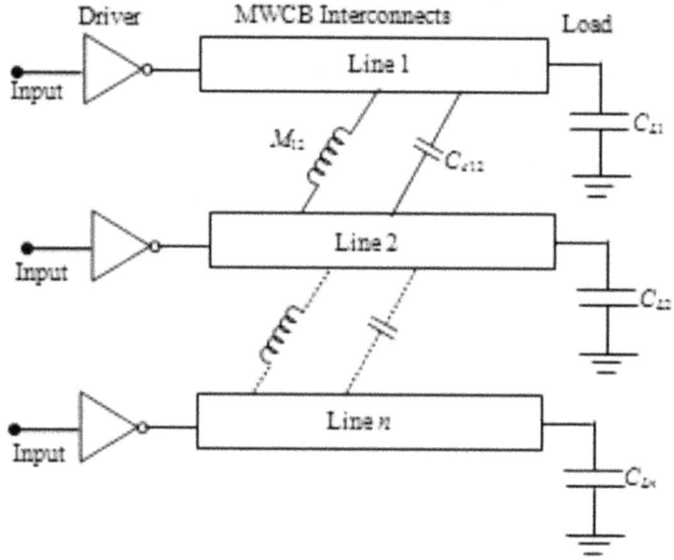

These interconnect lines experience a Miller coupling capacitive effect (C_c) depending on the spacing between them (Agrawal, 2016). The simulative analyses have been performed using HSPICE (HSPICE, 2016) in the present proposed work.

The performance analysis viz. delay and power dissipation of varying MWCB structures are shown in Figure 7 and 8 respectively. The on-chip wire length is varied from 500 to 2000μm. It is observed that power dissipation increases linearly while propagation delay has exponential rise with interconnect length. It is also inferred from figure the MWCB structure with maximum tube density i.e. MWCB-I possess least propagation delay and power dissipation at all interconnect lengths.

The performance improvement has been realized due to incorporation of high tube density MWCB structure. From (7) and (14), the increase in conducting channels enables reduction in scattering resistance while increasing quantum capacitance.

However, as seen in (12), the equivalent capacitance (C_{ESC}) is dominated by 1/ $C_{e,ESC}$ factor. Therefore, the cumulative effect of $R_{d,ESC}$ and C_{ESC} leads to an overall reduction of propagation delay and power dissipation in the interconnect structures under test. It is noticed from Fig. 7, at 2000μm of interconnect length, the reduction in delay for MWCB-I is 46.38%, 60.19% and 73.09% compared to that of MWCB-II, MWCB-III and MWCB-IV respectively. Fig. 8 shows the power dissipation increases with increase in interconnect length. For instance, in case of MWCB-I structure, the increase in power dissipation observed from 500 to 2000μm interconnect length is 19.38%.

Figure 7. Propagation delay varying with interconnect length for different MWCB structures

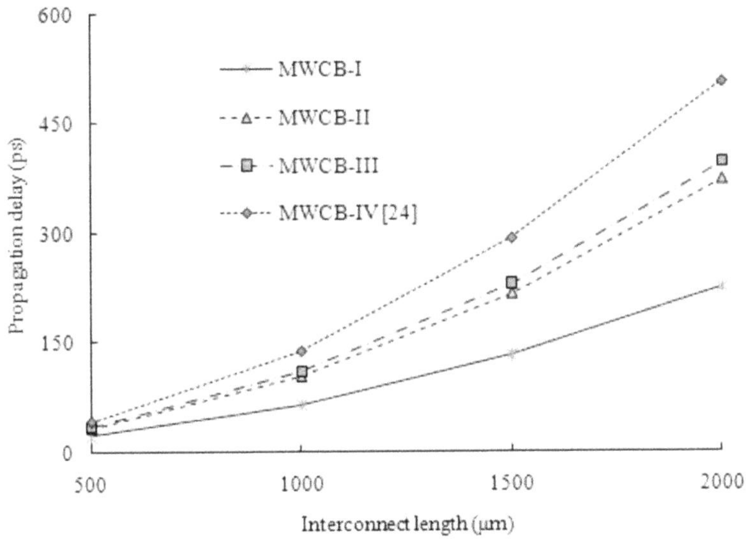

Figure 8. Power dissipation varying length of on-chip interconnect for different MWCB structures

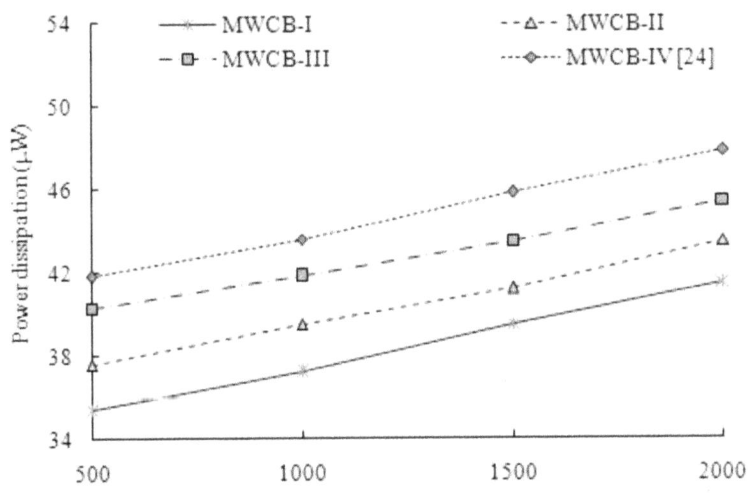

The worst-case crosstalk delay (out-phase) has been analyzed for all the four MWCB structures and is illustrated in figure 9. Crosstalk delay increases with increase in transition time period which relies on rise time and fall time. Under dynamic crosstalk effect, increase in transition time, is accompanied by an increase in the signal propagation delay (Kang, 2002). For instance, at an interconnect length of 1000μm and a time period of 20ps, crosstalk delay of MWCB-I reduces by 40.46%, 54.40% and 68.23% as compared to MWCB-II, MWCB-III and MWCB-IV respectively. When an input signal switches from high to low or vice versa, both the transistors in CMOS circuit conduct simultaneously from the time when ramp voltage is between V_{tn} to V_{dd} - $|V_{tp}|$ (Kang, 2002). This leads to short circuit current and subsequently short circuit power dissipation.

From figure 10, it is observed that power dissipation escalates with the rise in transition period. Further, it is seen that MWCB-I leads to least power dissipation compared to the other three structures under consideration. The overall average fall in power dissipation for MWCB-I structure is 18.97%, 21.44% and 27.37% compared to MWCB-II, MWCB-III and MWCB-IV respectively.

Furthermore, multi-line interconnect systems comprising of three and five number of coupled MWCBs respectively have been considered for crosstalk analysis in the present study. The middle interconnect is the victim line. The propagation delay computed for 3-Line coupled interconnect system is depicted in Table 4. In Case-(iii), inputs of aggressor line inputs switch in the same direction, while victim line switches in the opposite direction as that of aggressor lines. This represents out-phase switching. It is observed that, the propagation delay is 0.272ns, 0.594ns, 0.778ns and 1.108ns for MWCB-I, MWCB-II, MWCB-III and MWCB-IV respectively. The delay in case-(iii) is 4.25 times higher than that of case-(i) i.e. for MWCB-I interconnect. Such a high delay is attributed to comparatively higher Miller capacitive effect on the victim line in case-(iii).

Furthermore, Figure 11 presents the transient response at the victim line for varying MWCB structures in coupled 5 interconnect lines. Victim line 3 is in quiescent mode at logic high and the output for this line should be ideally at logic low. It is seen that the peak noise voltages in MWCB-I, MWCB-II, MWCB-III and MWCB-IV interconnects are 0.318V, 0.325V, 0.329V and 0.331V respectively. Thus, MWCB-I provides the lowest peak noise voltage.

Along with this, the ratios of propagation delay and power dissipation of copper to MWCB-I interconnect at varying minimized scaled nodes are evaluated. Ratio of copper to MWCB-I interconnect for propagation delay is shown in Figure 12. It is seen that propagation delay of copper interconnect is nearly 1.55 times and 2.1 times higher than MWCB-I interconnect for 70nm and 22nm deep submicron technologies respectively. This is due to lower parasitics associated with MWCB-I interconnect structure as counterpart to copper interconnect. The power dissipation

Figure 9. Crosstalk delay for varying interconnect length and transition period (a) MWCB-I (b) MWCB-II (c) MWCB-III (d) MWCB-IV

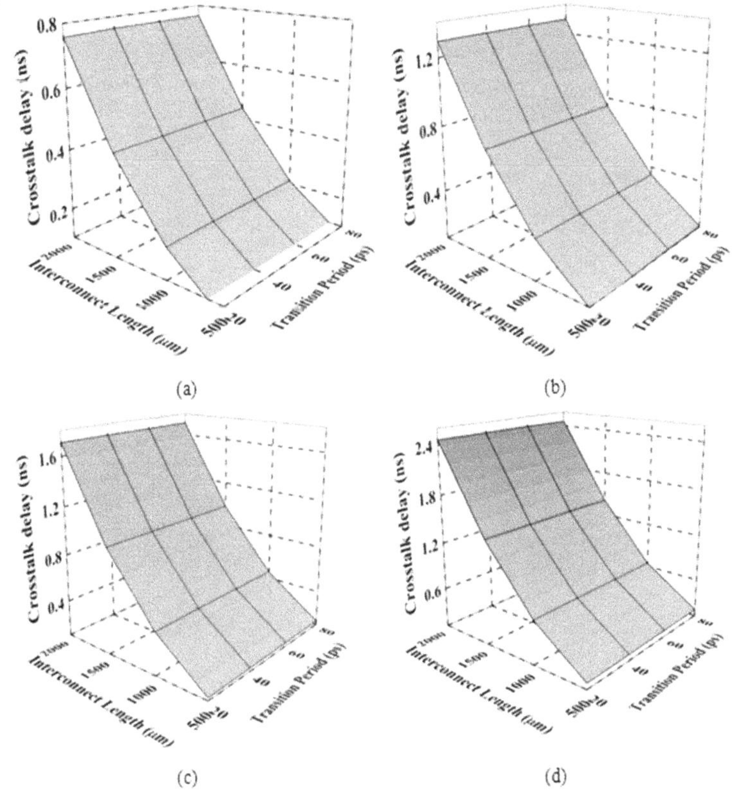

ratio of copper to the MWCB-I interconnect is seen in Figure 13. It is seen that the power dissipation of copper interconnect is 1.9 times and 1.58 times higher than that of MWCB-I interconnects for 90nm and 22nm technology nodes respectively.

It is therefore evident that the effect of functional crosstalk is least in MWCB-I interconnect as counterpart to the other three interconnect structures. It is inferred from the analysis that the performance of MWCB-I, in terms of crosstalk delay, power dissipation and propagation delay is superior amongst the four structures considered in the present study. Thus, the analysis verifies the efficacy of the CBO optimization technique utilized for the MWCB structures. In addition, the substantial decrement is noticed in terms of propagation delay and power dissipation of MWCB-I as compared to copper interconnect at nano dimensions.

Figure 10. Power dissipation for varying interconnect length and transition period (a) MWCB-I (b) MWCB-II (c) MWCB-III (d) MWCB-IV

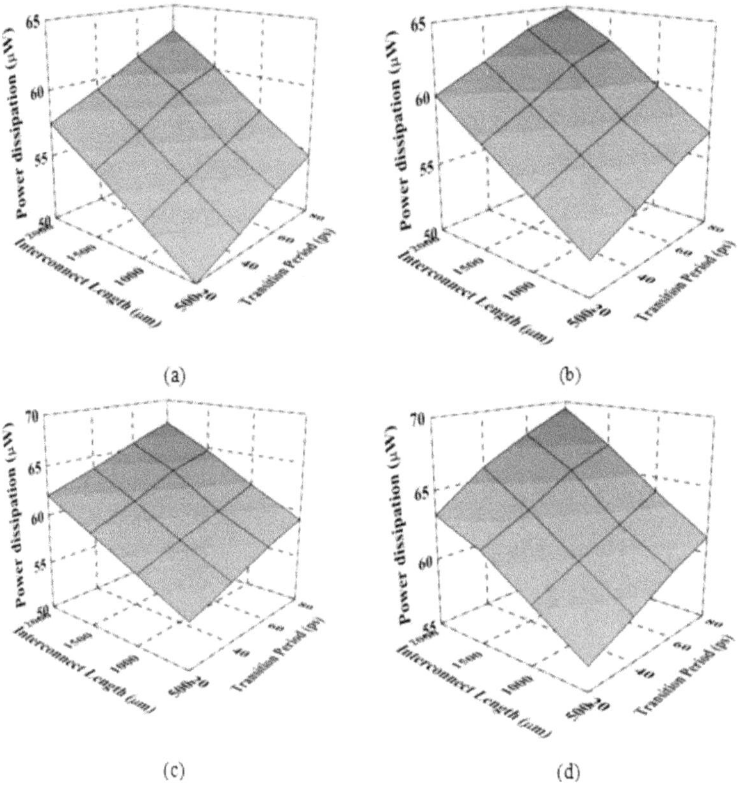

Table 4. Propagation delay analysis of 3-couples MWCB interconnect system

Case No.	Input Signal Switching Conditions			Propagation Delay (ps)			
	Aggressor	**Victim**	**Aggressor**	**MWCB-I**	**MWCB-II**	**MCWB-III**	**MCWB-IV (Majumder, 2015)**
Case-(i)	1→0	1→0	1→0	64.03	113.78	149.66	218.79
Case-(ii)	1→0	1→0	0→1	173.55	294.15	385.45	555.47
Case-(iii)	1→0	0→1	1→0	272.28	594.54	778.7	1108.2
Case-(iv)	1→0	0→1	0→1	171.24	291.95	383.1	551.01

Figure 11. Transient analysis for 5-coupled interconnect lines

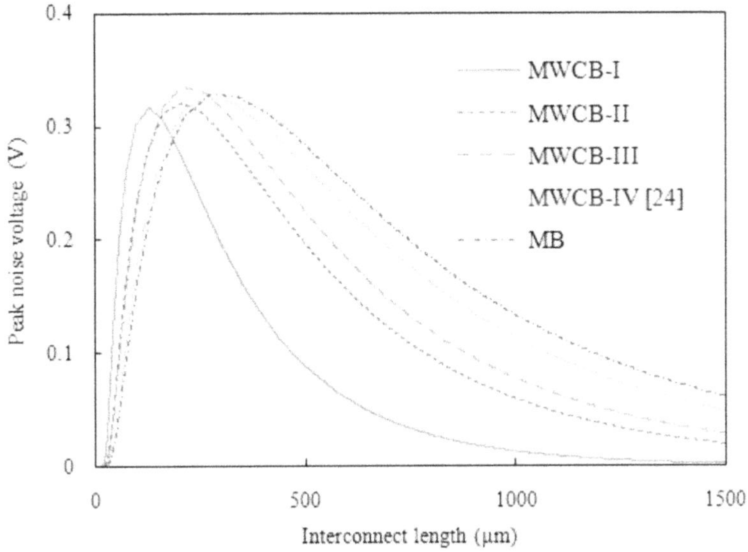

Figure 12. Propagation delay ratio of copper to MWCB-I interconnect for different technology nodes

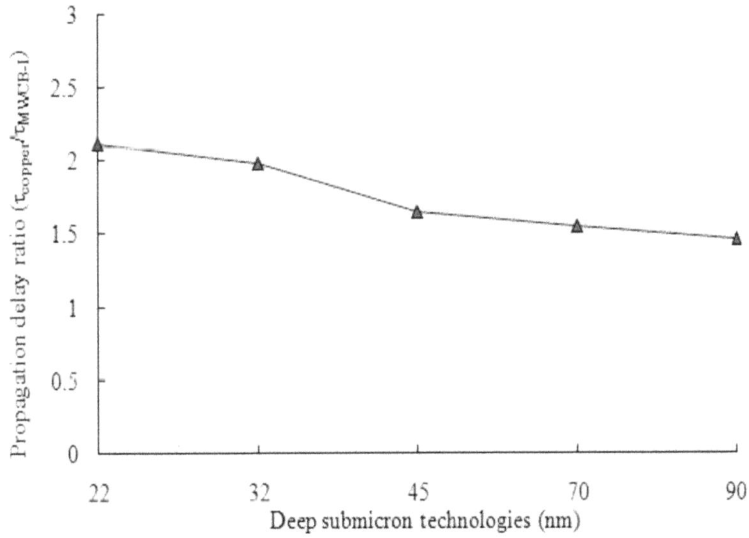

Figure 13. Power dissipation ratio of copper to MWCB-I interconnect for different technology nodes

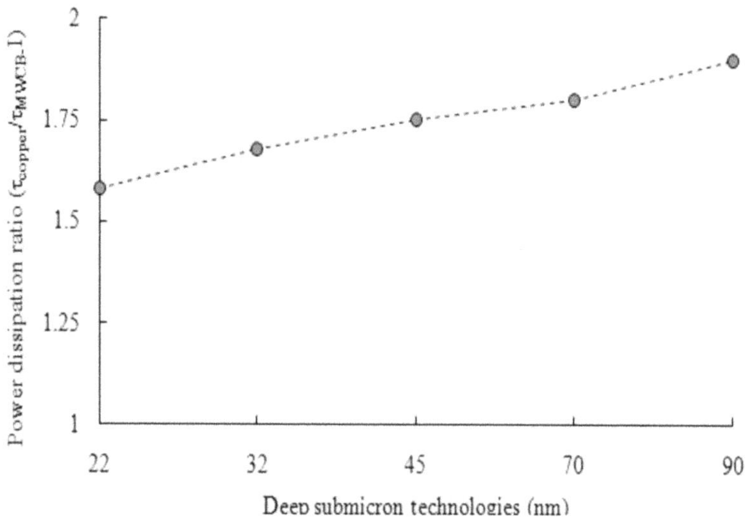

5. CONLCUSION

In this chapter, prospective and efficient colliding bodies optimization (CBO) technique is effectively incorporated to optimize mixed-wall CNT bundle (MWCB) structure. This is attained by maintain maximum tube density. The coupled interconnect structure is used to assess the performance of the system. Comparative analysis of propagation delay, crosstalk and power dissipation has been performed for MWCB interconnect structures with different tube densities. It is promisingly seen that the MWCB structure with higher tube density given lesser crosstalk, delay and power dissipation. Consequently, it can be positively inferred that on-chip interconnect structure with increased tube density can give high performance. The incorporated CBO technique is an effective method to attain high tube densities and enhanced performance of system.

REFERENCES

Agrawal, Y., & Chandel, R. (2016, July). Crosstalk analysis of current-mode signalling-coupled RLC interconnects using FDTD technique. *IETE Technical Review*, *33*(2), 148–159. doi:10.1080/02564602.2015.1056258

Agrawal, Y., Kumar, M. G., & Chandel, R. (2016, July). Comprehensive model for high-speed current-mode signaling in next generation MWCNT bundle interconnect using FDTD technique. *IEEE Transactions on Nanotechnology*, *15*(4), 590–598. doi:10.1109/TNANO.2016.2558475

Cheung, C. L., Kurtz, A., Park, H., & Lieber, C. M. (2002, February). Diameter-controlled synthesis of carbon nanotubes. *The Journal of Physical Chemistry B*, *106*(10), 2429–2433. doi:10.1021/jp0142278

Giustiniani, A., Tucci, V., & Zamboni, W. (2010, August). Modeling issues and performance analysis of high-speed interconnects based on a bundle of SWCNT. *IEEE Transactions on Electron Devices*, *57*(8), 1978–1986. doi:10.1109/TED.2010.2050836

Haruehanroengra, S., & Wang, W. (2007, August). Analyzing conductance of mixed carbon-nanotube bundles for interconnect applications. *IEEE Transactions on Electronic Devices Letters, 28*(8), 756-759.

Hone, J., Whitney, M., Piskoti, C., & Zettl, A. (1999, January). Thermal conductivity of single-walled carbon nanotubes. *Physical Review B: Condensed Matter*, *59*(4), R2514–R2516. doi:10.1103/PhysRevB.59.R2514

Huang, W. Q., Li, Y., Akeb, H., & Li, C. (2004, October). Greedy algorithms for packing unequal circles into a rectangular container. *The Journal of the Operational Research Society*, *56*(5), 539–548. doi:10.1057/palgrave.jors.2601836

Kang, S. M., & Leblebici, Y. (2002). *CMOS Digital Integrated Circuits Analysis and Design*. New York: McGraw-Hill.

Kaveh, A., & Mahdavi, V. R. (2014, May). Colliding bodies optimization: A novel meta-heuristic method. *Computers & Structures*, *139*(1), 18–27. doi:10.1016/j.compstruc.2014.04.005

Kaveh, A., & Mahdvi, V. R. (2014, January). Colliding bodies optimization method for optimum design of truss structures with continuous variables. *Advances in Engineering Software*, *70*(1), 1–12. doi:10.1016/j.advengsoft.2014.01.002

Koo, K. H., Cho, H., Kumar, P., & Saraswat, K. C. (2007, October). Performance comparisons between carbon nanotubes, optical, and Cu for future high-performance on-chip interconnect applications. *IEEE Transactions on Electron Devices*, *54*(12), 3206–3215. doi:10.1109/TED.2007.909045

Kumar, M. G., Agrawal, Y., & Chandel, R. (2016, April). Carbon nanotube interconnects- a promising solution for VLSI circuits. *IETE Journal of Education, 57*(2), 46–64. doi:10.1080/09747338.2016.1158129

Kumar, V. R., Kaushik, B. K., & Patnaik, A. (2015, December). Improved crosstalk noise modeling of MWCNT interconnects using FDTD technique. *Microelectronics Journal, 46*(12), 1263–1268. doi:10.1016/j.mejo.2015.10.009

Li, H., & Banerjee, K. (2009, September). Carbon nanomaterials for next-generation interconnects and passives: Physics, status, and prospects. *IEEE Transactions on Electron Devices, 56*(9), 1799–1821. doi:10.1109/TED.2009.2026524

Majumder, M.K., Das, P.K., Kumar, V.R., & Kaushik, B.K. (2015, August). Crosstalk induced delay analysis of randomly distributed mixed CNT bundle interconnect. *Journal of Circuits, Systems and Computers, 24*(10), 1550145-1- 1550145-15.

Majumder, M. K., Kumar, J., Kumar, V. R., & Kaushik, B. K. (2014, October). Performance analysis for randomly distributed mixed carbon nanotube bundle interconnects. *Micro & Nano Letters, 9*(11), 792–796. doi:10.1049/mnl.2014.0502

Novoselov, K. S., Geim, A. K., Morozov, S. V., Jiang, D., Zhang, Y., Dubonos, S. V., ... Firsov, A. A. (2004, October). Electric field effect in atomically thin carbon films. *Science, 306*(5696), 666–669. doi:10.1126cience.1102896 PMID:15499015

Panda, A., & Pani, S. (2014, December). *A New model based on colliding bodies optimization for identification of Hammerstein Plant.* Paper presented at IEEE INDICON, Pune, India. 10.1109/INDICON.2014.7030381

Pu, S. N., Yin, W. Y., Mao, J. F., & Liu, Q. H. (2009, April). Crosstalk prediction of single- and double-walled carbon-nanotube (SWCNT/DWCNT) bundle interconnects. *IEEE Transactions on Electron Devices, 56*(4), 560–568. doi:10.1109/TED.2009.2014429

Rao, S. S. (1984). *Optimization: Theory and Applications*. New York: Wiley.

Sarto, M. S., & Tamburrano, A. (2010, January). Single-conductor transmission line model of multiwall carbon nanotubes. *IEEE Transactions on Nanotechnology, 9*(1), 82–92. doi:10.1109/TNANO.2009.2023516

Semiconductor Industry Association, International Technology Roadmap for Semiconductors (ITRS). (2012). Available: http://public.itrs.net

Srivastava, A., Xu, Y., & Sharma, A. K. (2010, April). Carbon nanotube for next generation very large scale integration interconnects. Journal of Nanophotonics. *Society of Photo-Optical Instrumentation Engineers, 4*(1), 041690-1–041690-26.

Srivastava, N., Li, H., Kreupl, F., & Banerjee, K. (2009, July). On the applicability of single-walled carbon nanotubes as VLSI interconnects. *IEEE Transactions on Nanotechnology*, *8*(4), 542–559. doi:10.1109/TNANO.2009.2013945

Synopsys for HSPICE tools. (2016). Available: http://www.synopsys.com

Wang, W., Haruehanroengra, S., Shang, L., & Liu, M. (2007, April). Inductance of mixed carbon nanotube bundles. *Micro & Nano Letters*, *2*(2), 35–39. doi:10.1049/mnl:20070027

Wei, B. Q., Vajtai, R., & Ajayan, P. M. (2001, August). Reliability and current carrying capacity of carbon nanotubes. *Applied Physics Letters*, *79*(8), 1172–1174. doi:10.1063/1.1396632

Yilmazoglu, O., Joshi, R., Popp, A., Pavlidis, D., & Schneider, J. J. (2011, January). Pronounced field emission from vertically aligned carbon nanotubes blocks and bundles. *Journal of Vacuum Science & Technology B Microelectronics and Nanometer Structures*, *29*(2), 02B106-1–02B106-5.

Yu, M. F., Lourie, O., Dyer, M. J., Moloni, K., Kelly, T. F., & Ruoff, R. S. (2000, January). Strength and breaking mechanism of multiwalled carbon nanotubes under tensile load. *Science*, *287*(5453), 637–640. doi:10.1126cience.287.5453.637 PMID:10649994

Zhu, L., Xu, J., Xiu, Y., Sun, Y., Hess, D. W., & Wong, C. P. (2006, February). Growth and electrical characterization of high-aspect-ratio carbon nanotube arrays. *Carbon*, *44*(2), 253–258. doi:10.1016/j.carbon.2005.07.037

Section 4
Future of CNTFET Technology

Chapter 10
Transport Properties of Silicene Nanotube- and Silicene Nanoribbon-Based FETs

Deep Kamal Kaur Randhawa
Guru Nanak Dev University, Regional Campus, Jalandhar, India

Paramjot Singh
https://orcid.org/0000-0003-3830-9583
Guru Nanak Dev University, Regional Campus, Jalandhar, India

Tarun
Guru Nanak Dev University, Regional Campus, Jalandhar, India

ABSTRACT

Silicene is one of the most interesting nanomaterials. In this chapter, computational studies have been done on Silicene nanotube and nanoribbon-based FETs to analyze their transport properties. The FET is designed from armchair nanoribbon and single wall nanotube. The scattering region is capped by a dielectric and a metallic layer to form a gate. The conductance versus gate bias voltage, conductance versus temperature up to 2000K, and electrode temperature versus current characteristics are calculated and plotted along with the design of the equivalent model of the structure. Extended Huckel-based calculations were used, and the analysis shows the transport properties of both structures.

DOI: 10.4018/978-1-7998-1393-4.ch010

1.1 INTRODUCTION

Electrical and electronic devices initially comprised of two terminal devices, in which current flowed from one terminal to the other terminal. With development of technology three terminal devices were devised which we could control the flow of current between two terminals by the third terminal. Under category of three terminals devices, two models are widely accepted that is, Bipolar Function Transistor (BJT) and Field Effect Transistor (FET). FET is more widely used in Integrated circuits (ICs) as compare to BJT virtue its better electronic characteristics. The introduction of Field effect Transistors (FET) in the field of electronics has revolutionized the industry as the second revolution in electronics was introduced by the introduction of MOSFET's in high speed desktops and hand held calculators.

A FET is a three-terminal device that uses electric field to control the flow of current. The three terminals are called source, gate and drain. Gate is applied with external voltage to control the current flowing from source to drain. FET is a unipolar device as it operates with the flow of only one type of charge carriers. As the field effect transistor the width of the channel along with the current can be varied with the external electrical field, hence FET is a voltage controlled device. FETs can N-type or P-type channel device.

A FET has three basic terminals. Source: the terminal from which the carrier enters (current denoted by I_S). Drain: the terminal from which the carrier leaves the channel (current denoted by I_D). Gate: the terminal that modulates the entire channel by applying voltage (current denoted by I_G) (sedra, 2003). All the terminals are attached with each other with ohmic contacts. FET can be classified into two types Metal-oxide semiconductor field effect transistor and junction field effect transistors. The basic working principle of a transistor is explained in figure 1.

Figure 1. Basic principle of a FET device

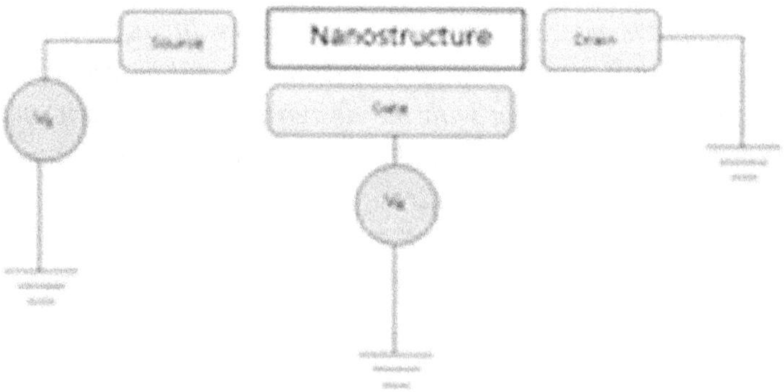

The current through the channel from source to drain is controlled through gate by applying the external voltage. The FET's have their applications in various integrated circuits and as high performance switches the present of metal oxide or dielectric layers FET exhibits high input impedance which makes them more sensitive towards the input signal. Miniaturization of devices in pursuit of faster and portable devices has been the trend of electronics leading to more complex circuits in effect enhancing the performance. This has also made high density memories a reality. As the device size is moving to nanometer regime the circuits and memories will be including nano-scaled transistors, maybe a single molecule transistor that can store data.

The ever-growing electronics industry is based on roadmap predicted by Gordon Moore which states that the numbers of transistors on an integrated circuit will double every year. This roadmap is stated as Moore's law. This development is pushing electronics into the realm of nanotechnology and in effect approaching silicon wall. So there are limitations on silicon based transistors. The use of nanomaterials in the transistor fabrication offer a very realistic alternate. So there is need to analyze nanomaterials for their electronic properties so as to design Nano-FETs.

1.2 NANOMATERIALS

Nanotechnology is the branch of science which deals with material that lies in nano-scale range (1-100) Nm. Nanomaterials are defined as materials that have at least one of its dimensions in the nanoscale range. These materials are classified into three basic types according to their dimensions which is one dimensional, two dimensional and three-dimensional nanomaterials. The one-dimensional nanomaterial has only one of its dimensions out of nanoscale range, for example nano-rods. Two dimensional nanomaterials have two dimensional out of nanoscale range such as nanosheets. The three-dimensional nanostructures are bulk materials made up from the assembly of individual nanoparticles such as bulk powders and bundle of nanowires. The term nanotechnology is not confined to any one branch as the nanotechnology has its application in a variety of fields such as chemistry, physics, health and medicine, electronics and many others. The introduction of this science in the field of electronics has led to the development and enhancement of various devices. The ability of particles to change their physical and chemical properties as their size changes in nanoscale makes the nanotechnology a hot topic these days. Due to the large surface area to volume ratio, the physical and chemical properties vary when the size of the material is changed in nanoscale. The significance of nanoscience in electronics is to enhance the properties of the device by fabricating it in the nanoscale range which will ultimately improve its power consumption and cost. Sensitivity and selectivity are the two main factors leading the research to be carried out so as to find efficient devices. The term nanoelectronics covers a diverse set of devices and materials that are so small that their atomic interactions and quantum mechanical properties need to be studied intensively. Nanoelectronics will provide a viable solution to sustain Moore's Law.

1.2.1 Nanoribbons

Nanoribbon is a two-dimensional structure that exhibits distinct properties which has a variety of applications in different fields. With the advancement in nano-electronics, nanoribbons have been investigated for their potential as a nano-sensors also. Materials such as graphene, silicene and germanene have shown excellent sensing capabilities towards various gases and molecules. This behavior has shown us the importance of two-dimensional nanoribbons in the field of sensors. Nanoribbons are generally categorized into two types Armchair and Zig-zag. Following are some common nanoribbons that are current being studied for their properties.

1.2.1.1 Graphene

Graphene is the single atomic thick solitary layer of graphite. This two-dimensional structure exhibits exceptional electrical, physical and thermal properties. Graphene is a hexagonal honeycomb single atom thick layer structure in which carbon atoms are compactly packed with sp^2 hybridization. The C-C bond length in graphene sheet is 1.42Å and it contributes to form the most stable allotrope of carbon which is Graphite. The average spacing between the adjacent graphene layers in graphite is 0.335Å which is due to Van der Waal interactions. It is also the fundamental element of other carbon allotropes like fullerenes, charcoal and nanotubes. Graphene is a zero bandgap semiconductor material that offers high mobility at room temperature. The density and breaking strength of graphene is 0.77 mg/m^2 and 42 N/m respectively. The availability of a free electron in the third dimension of the graphene sheet provides it with the property of high electric conductivity. The thermal conductivity of graphene is about 5000 $Wm^{-1}K^{-1}$ compared to copper which is 401 $Wm^{-1}K^{-1}$ (Yang,2010; Huang,2007; Romero, 2009; Kyozaburo, 1994). Recent studies have shown that graphene has a promising behavior towards sensing due to its distinct properties.

1.2.1.2 Silicene

Silicene is the analogue of graphene made up from single atom thick layer of silicon. The structure of silicene is hexagonal honeycomb similar to graphene. Silicene was found out and studied in 1994 by Takeda and Shiraishi. It was named by Guzman-Verri and Voon in 2007. Silicene was self-assembled on silver (Ag) with miller indices (111) and SEM images confirmed that silicene has a hexagonal structure similar to graphene. The silicene has properties like ferromagnetism (Wang, 2012), superconductivity (Chen, 2013), half-metallicity (Zheng, 2012) and quantum hall effect (Liu, 2011). The silicon possesses both sp^2 and sp^3 hybridization which makes silicene a buckled structure at its possible lowest energy levels unlike graphene, which is planer sp^2 hybridized (Zaminpayma, 2016; Tsai, 2013). The band gap tuning can be done more intensively with an external electric field as, the buckled structure is more stable compared to the planer. The Si-Si bond length in silicene is 2.23Å with a buckling height of 0.44Å. Owing the above mentioned properties silicene have been proposed for gas sensing. Studies have reviled that silicene has excellent sensing capabilities towards gas molecules (Walia, 2018; Aghaie, 2016; Akbari, 2016; Balendhran, 2014; Walia, 2018; Hu, 2014). Silicene is also a zero bandgap material however with the introduction of defects and dopants the bandgap can be changed for analyzing its properties. A silicene nanoribbon is shown in the figure 3 with different bondlengths.

Figure 2. Graphene nanoribbon

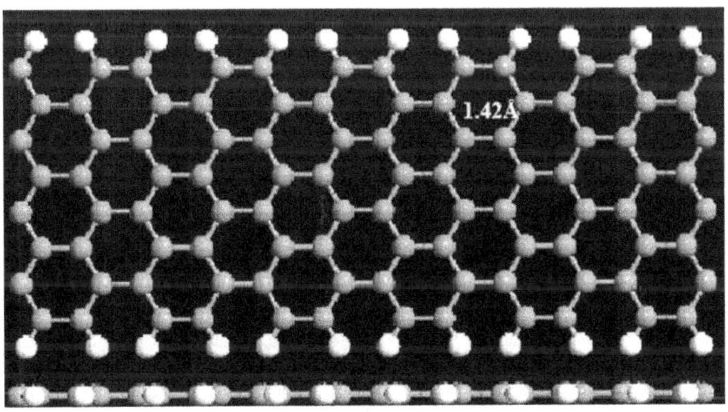

1.2.1.3 Germanene

Germanene is a single atom thick layer structure made up from germanium atoms. Similar to silicene, it has a hexagonal honeycomb structure. The Ge-Ge bond length is about 2.44 Å with a buckling height of 0.68 Å. Germanene was first reported by G. Le Lay as the deposition of single atom thickness order and two-dimensional structure of Ge atoms film by Molecular-Beam Epitaxy upon a Gold surface in a crystal lattice. STM confirmed nearly flat germanene but the stable buckled structure was proposed afterward. The electronic and optical properties of germanene make the structure useful for high-performance field effect transistors (FET's). DFT and first principles investigation have shown that germanene can be used as a sensor for gases and bio-molecules. It is optically active material hence it has much application in the field of optoelectronics. Exceptional spin-orbit coupling of germanene provides the researchers to study quantum spin Hall Effect at room temperature. Analysis has shown that germanene can be used as gas sensor and in this study adsorption behavior of germanene is studied for Uric acid (Davila, 2016).

1.2.2 Nanotubes

Nanotube is one of the most useful nanostructure formed by the physicist and material scientist. The most generalized method for its preparation is the rolling up of the nanosheet or nanoribbon. We can roll the nanosheet or nanribbon at different angle and accordingly the characteristic behavior of the nanotube also changes.

Figure 3. Silicene nanoribbon

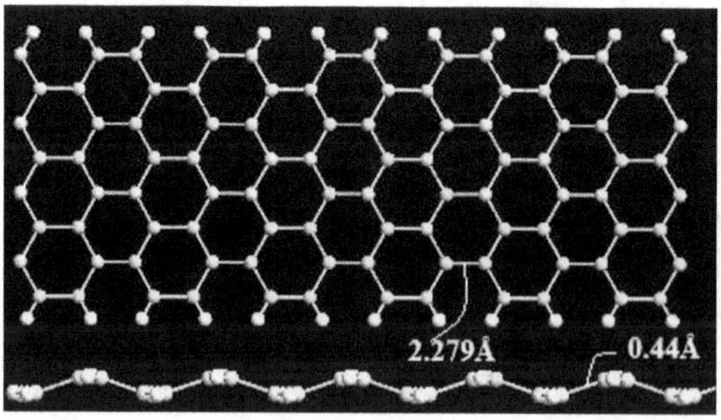

Figure 4. Germanene nanoribbon

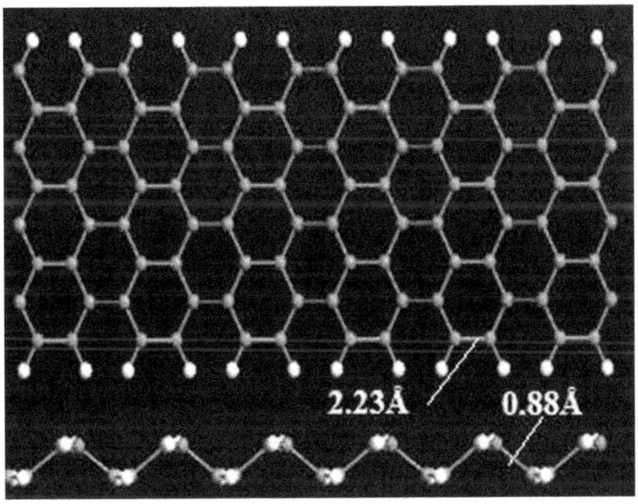

Double wall Nanotubes are those nanotubes in which their layer of another nanotube inside main nanotube. Whereas, in single wall nanotube there is only nanotube. Multi wall Nanotubes are those nanotubes in which their number of layers of other nanotubes inside main nanotube shown in Figure 5.

1.2.2.1 Carbon Nanotube

Carbon nanotubes are the nanomaterials formed by two-dimensional hexagonal lattice of carbon atoms. The carbon atoms sheet graphene is bent and joined in a direction so as to form a hollow cylinder. CNT's are a class of carbon allotropes that carry exceptional electric and transport properties. The Orbital hybridization in CNT results in sp^2 hybridization in carbon atoms which is stronger than sp^3 hybridization in diamond.

Figure 5. (a)Single-wall carbon nanotube (b)Double-wall carbon nanotube. (c) Multi-wall carbon nanotube (Michele, 1970)

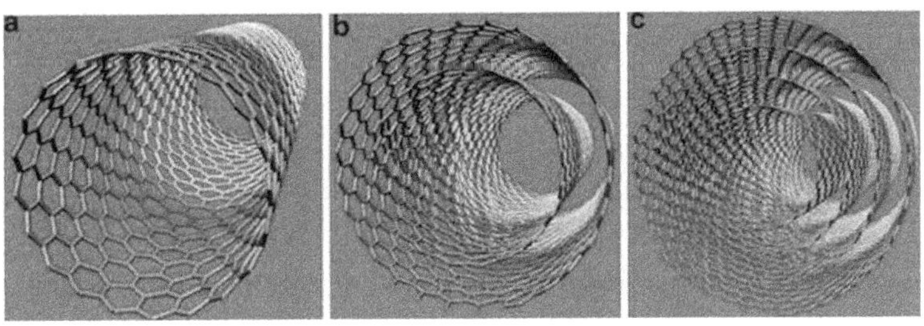

Carbon nanotubes are basically of two types single wall CNT or Multi wall CNT. The multi wall CNT has multiple graphene sheets rolled over one other. Carbon nanotubes have thermal, electrical and transport properties hence they are prominent materials for advance electronic devices. Similar to graphene CNT's are either metallic or semi-metallic depending on the direction in which sheet is rolled. SWCNT with small diameters are strong as they have high Young's modulus and high tensile strength. Similar to carbon nanotubes we have silicene nanotube made up by rolling silicene sheets (Kima, 2012).

1.2.2.2 Silicene Nanotube

Silicene is analogous to graphene, a two-dimensional nanostructure made up of silicon atoms. The silicene has a hexagonal honeycomb structure similar to graphene. Silicene can also be used to make nanotubes called silicene nanotubes. Though, silicene nanotubes have not been experimentally extracted but theoretical studies have shown that silicene has properties somewhat similar to graphene. The silicene nantube is created similar to CNT's by rolling silicene nanosheets in a desired direction. The Si-Si bond length in SNT is same as silicene 2.279Å. from the bulk silicene nanotube the three probe model is constructed to create silicene nanotube FET device. Two electrodes and a scattering region are created on SNT and a metallic layer is capped over the gate. Gate here is the central region from which the channel width and current can be controlled using the external voltage.

1.3 SILICENE BASED NANOFETS

The term nano FET refers to the FET device that has its dimensions in nano-scale range, the FET can be a single atom or multi atom depending upon the size from 1 to 100nm. Computational studies have shown that nanomaterials have exceptional properties that changes with change in its size and chirality (Fiori, 2012; Zare, 2013; Ouyang, 2007). Carbon nanotubes, graphene and many other materials have been studied for their properties as electronics devices. Various computational applications provide the platform where one can calculate and analyze the transport properties of the nanodevices. One of the software is Virtual Nano Lab Atomistix Tool Kit (VNL-ATK). In this analysis a silicene nanotube has been taken that is basically a rolled sheet of two dimensional silicene naoribbon and the transport properties of silicene nanotubes as Field effect transistor are calculated. All calculations are carried out on simulation package Virtual Nano Lab-Atomistix Tool Kit (VNL-ATK) based on Extended huckel theory. Armchair Silicene Nanoribbon (ASiNR) 6 atoms wide is taken. The Si-Si bond length is 2.279Å and the structure is edge passivated by hydrogen atoms as edges are more reactive than sides. The electron temperature and density mesh cut-off is set at 300K and 10 Hartree respectively. The K point sampling of 1×1×100 is set for transport calculations.

Figure 6 shows silicene nanoribbon FET structure that has three terminals similar to a normal FET that is source, gate and drain. The gate or the scattering region has a metallic layer caped in cylindrical frame. The main region is the central region which is called the scattering region. The current through the channel can be controlled by external voltage applied at gate.

Similar to the nanoribbon a nanosheet is constructed and rolled to form a silicene nanotube. The metallic layer is same as in case of the nanoribbon. The only difference is that the sheet is rolled which have effect on change in transport properties.

Figure 6. Silicene nanoribbon FET with metallic coverage which act as gate for device

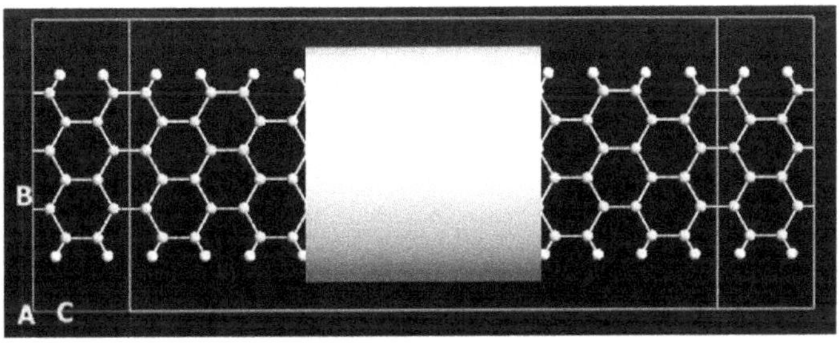

Figure 7. Silicene nanotube FET with metallic coverage which act as gate for device

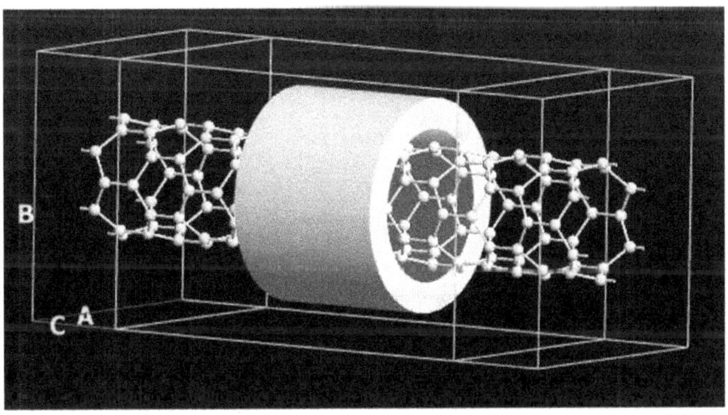

1.4 TRANSPORT PROPERTIES OF SILICENE BASED FETS

All the Transport properties are calculated using the circuitry shown in figure 8. Where V_{GS} is the voltage between the gate and the source, V_{DS} is the voltage between the source and the drain.

1.4.1 Silicene Based Nanoribbon FET

Conductance V/S Gate Voltage

The conductance vs. the gate voltage for the nanoribbon FET is shown in figure 9 the value of conductance was around 10^{-12} S at zero external bias. With the increase in the voltage the conductance increased until 2 V which was around 10^{-10} S after further increase in voltage the conductance decreased to 10^{-11} S until 3V and remained almost constant afterward. The external gate voltage is used to control the width and current carrying capability of FET. From the graph it is clear that with the change in the voltage the conductance is changing hence the current is changing. This shows that the silicene based FET's channel width and current carrying capability is varied with the external bias voltage

Conductance V/S Temperature

Figure 9 shows the conductance vs. the temperature of nanoribbon FET, the value of the conductance increased as the temperature in increased from 0K to 2000K. the increase is not linear the curve shows as, we further increases the temperature the conductance becomes stable after some critical temperature is attained.

Figure 8. The Schematic of Silicene based FET

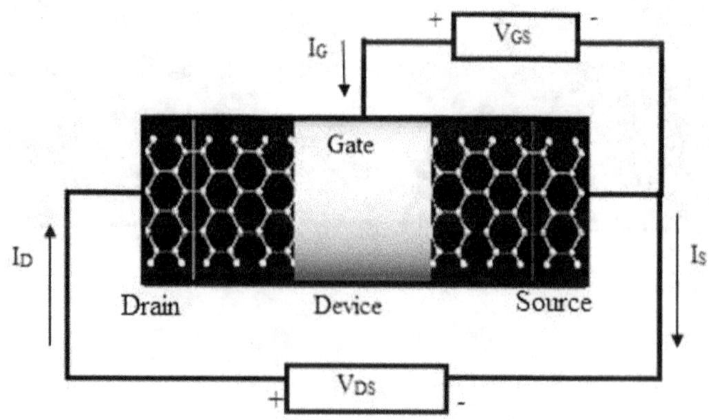

Figure 9. Conductance vs. the gate voltage, Conductance vs. temperature and Current vs. electrode temperature of nanoribbon FET

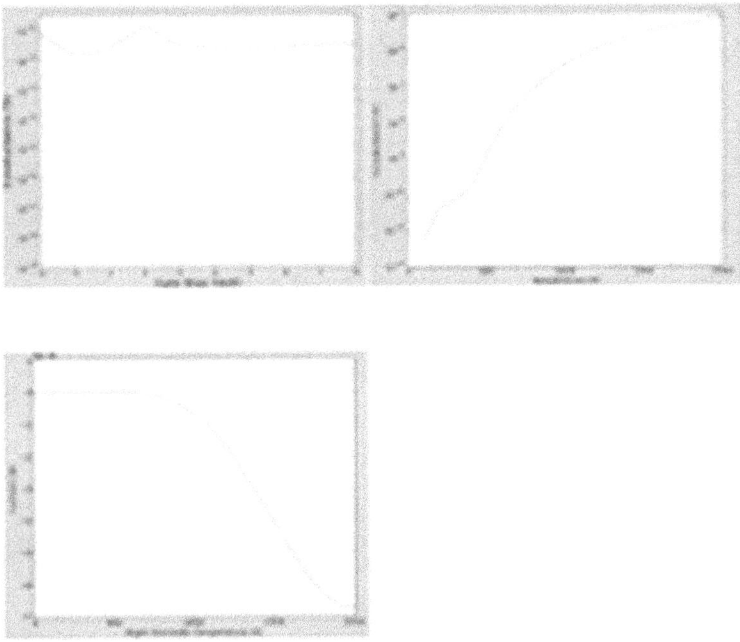

Current V/S. Electrode Temperature

Figure also shows the Current vs Electrode temperature graph of nanoribbon FET the value of current decreases with increase in electrode temperature. The current remained constant until 1000K and then started decreasing linearly.

Conductance increased as the temperature in increased from 0K to 2000K. the increase is not linear the curve shows as, we further increases the temperature the conductance becomes stable after some critical temperature is attained.

1.4.2 Silicene Based Nano Tube FET

Conductance V/S Gate Voltage

For the nanotube figure 10 shows the conductance versus Gate voltage graph, the variation is very less as the voltage is changed. The conductance increased linearly until the voltage is zero then decreased linearly afterwards. This shows that the conductance changes when the external gate voltage is varied hence the current through channel can be controlled through the external gate voltage.

Figure 10. Conductance Vs. gate voltage, Conductance vs. temperature and Current vs. electrode temperature of nanotube FET

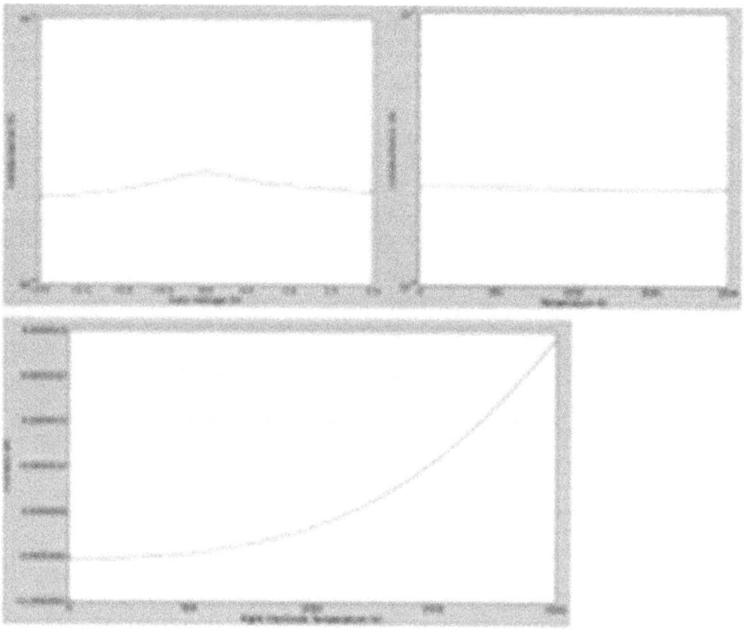

Conductance V/S Temperature

Figure 10 also shows the Conductance vs temperature graph of nanotube FET, the conductnce remain almost constant throughout the temperature variation . this means that the conductance here is indipendent of the temperature. Hence the conductance does not depends upon the tempertaure in case of nanotube FET

Current V/S. Electrode Temperature

The Current vs. electrode temperature of nanotube FET exhibits that the current increased as the electrode temperature is increased. The current reaches a max value of 2.5 μA at electrode temperature 2000 K

1.5 ADVANTAGES OF SILICENE FET

Silicene exhibits properties like ferromagnetism, Quantum Hall effect, superconductivity and half metallicity which provide the structure to serve as an electronic device such as field effect transistor. Spin polarization in silicene has its advantage over the other 2D structures for analyzing the electronic and transport properties. Silicene behaves as semi-metal as, the band gap tuning is possible with the external electric field hence it can have variety of applications accordingly in electronic devices

1.6 SUMMARY

The computational analysis carried out exhibit the transport properties of silicene nanoribon FET and silicone nanotube FET device. The analysis includes study of conductance, current, electrode temperature and gate bias. It is observed that the conductance varies as we increases the gate bias using external voltage. Further at higher electrode temperatures the current reduces as the electrode temperature increases. The conductance of the device also changes with the external temperature; as the temperature rises the conductance also increases. The devices are observed to be stable at very high temperatures. This shows that silicene material can be potentially used in nano field effect transistors.

REFERENCES

Aghaie, Monshi, & Calizo. (2016). *A Theoretical Study of Gas Adsorption on Silicene Nanoribons And Its Application as a Highly Selective Molecular Sensor.* Royal Society of Chemistry.

Akbari, Buntat, Afroozeh, Pourmand, Farhang, & Sanaati. (2016). *Slilicene and Graphene Nanomaterials in Gas Sensing Mechanism.* Royal Society of Chemistry.

Balendhran, Walia, Nili, & Sriram. (2014). *Elemental analogues of graphene: Silicene, Germanene, Stanene and Phosphorene*. Wiley.

Bonaccorso, F. (2010). Graphene photonics and optoelectronics. *Nature Photonics*. doi:10.1038/nphoton.2010.186

Chen, L., Feng, B., & Wu, K. (2013). Observation of a possible superconducting gap in silicene on Ag(111) surface. *Applied Physics Letters*, *102*(8), 081602. doi:10.1063/1.4793998

Dávila, M. E., & Le Lay, G. (2016). Few layer epitaxial germanene: A novel two-dimensional material Dirac material. *Scientific Reports*, *6*(1), 20714. doi:10.1038rep20714 PMID:26860590

Fiori, G., & Iannaccone, G. (2007). Simulation of Graphene Nanoribbon Field-Effect Transistors. *IEEE Electron Device Letters*. doi:10.1109/LED.2007.901680

Giulianini & Motta. (1970). *Polymer Self-assembly on Carbon Nanotubes*. doi:10.1007/978-1-4614-0742-3_1

Guzmán-Verr, G. G., & Lew Yan Voon, L. C. (2007). Electronic structure of silicon-based nanostructures. *Physical Review B: Condensed Matter and Materials Physics*, *76*(7), 075131. doi:10.1103/PhysRevB.76.075131

Hu, W., Xia, N., Wu, X., Liab, Z., & Yang, J. (2014). Silicene as a highly sensitive molecule sensor for NH3, NO and NO2. *Physical Chemistry Chemical Physics*.

Huang, B. (2007). Making a field effect transistor on a single graphene nanoribbon by selective doping. *Applied Physics Letters*. doi:10.1063/1.2826547

Kima, B., Leea, J., Namgunga, S., & Kima, J. (2012). DNA sensors based on CNT-FET with floating electrodes. *Sensors and Actuators. B, Chemical*, *169*, 182–187. doi:10.1016/j.snb.2012.04.063

Liu, C.-C., Feng, W., & Yao, Y. (2011). *Quantum Spin Hall Effect in Silicene and Two-Dimensional Germanium*. American Physical Society.

Ouyang, Yoon, & Guo. (2007). Scaling Behaviors of Graphene Nanoribbon FETs: A Three-Dimensional Quantum Simulation Study. *IEEE Transaction on Electronic Devices*.

Romero. (2009). Adsorption of ammonia on graphene. *Nanotechnology*.

Takeda, K., & Shiraishi, K. (1994). Theoretical possibility of stage corrugation in Si and Ge analogs of graphite. *Physical Review B: Condensed Matter and Materials Physics*. doi:10.1103/PhysRevB.50.14916

Tsai, Huang, Chang, Lin, Jeng, & Bansil. (2013). *Gated silicene as a tunable source of nearly 100% spin-polarized electrons*. NCBI.

Walia, G. K., & Deep, K. K. R. (2018). *Gas-sensing properties of Armchair silicene ASiNRs towards carbon-based gases with single-molecule resolution. In Structural Chemistry*. Springer.

Walia, G. K., & Deep, K. K. R. (2014). First-principles investigation on defect-induced silicene ASiNRs — A superior media for sensing NH3, NO2 and NO gas molecules. In Surface Science. Elsevier.

Walia & Randhawa. (2018). Adsorption and dissociation of sulfur-based toxic gas molecules on silicene ASiNRs: a quest for high-performance gas sensors and catalysts. *Journal of Molecular Modelling*.

Wang, X.-Q., Li, H.-D., & Wang, J.-T. (2012). Induced ferromagnetism in one-side semihydrogenated silicene and germanene. *Physical Chemistry Chemical Physics*.

Yang, X. (2010). Triple-mode single-transistor on a single graphene nanoribbon by selective doping. *Nano*.

Zaminpayma, E., & Nayebi, P. (2016). Band gap engineering in silicene: A theoretical study of density functional tight-binding theory. *Physica E, Low-Dimensional Systems and Nanostructures*. doi:10.1016/j.physe.2016.06.016

Zare, Shadmani, & Pournamdari. (2013). DFT/NBO study of Nanotube and Calixarene with anti-cancer drug. *Journal of Nanostructure in Chemistry*.

Zheng & Zhang. (2012). The electronic and magnetic properties of functionalized silicene: a first-principles study. *Nanoscale Res Lett., 7*(1), 422.

Compilation of References

Aghaie, Monshi, & Calizo. (2016). *A Theoretical Study of Gas Adsorption on Silicene Nanoribons And Its Application as a Highly Selective Molecular Sensor*. Royal Society of Chemistry.

Agrawal, Y., & Chandel, R. (2016, July). Crosstalk analysis of current-mode signalling-coupled RLC interconnects using FDTD technique. *IETE Technical Review*, *33*(2), 148–159. doi:10.1080/02564602.2015.1056258

Agrawal, Y., Kumar, M. G., & Chandel, R. (2016, July). Comprehensive model for high-speed current-mode signaling in next generation MWCNT bundle interconnect using FDTD technique. *IEEE Transactions on Nanotechnology*, *15*(4), 590–598. doi:10.1109/TNANO.2016.2558475

Akbari, Buntat, Afroozeh, Pourmand, Farhang, & Sanaati. (2016). *Slilicene and Graphene Nanomaterials in Gas Sensing Mechanism*. Royal Society of Chemistry.

Akturk, G., Pennington, G., Goldsman, N., & Wickenden, A. (2007). Electron transport and velocity oscillations in a carbon nanotube. *IEEE Transactions on Nanotechnology*, *6*(4), 469–474. doi:10.1109/TNANO.2007.896968

Alam, K., & Lake, R. (2007). Role of doping in carbon nanotube transistors with source/drain underlaps. *IEEE Transactions on Nanotechnology*, *6*(6), 652–658. doi:10.1109/TNANO.2007.908170

Ali, A., Khan, A., Karimov, K. S., Ali, A., & Daud Khan, A. (2018). Pressure sensitive sensors based on carbon nanotubes, graphene, and its composites. *Journal of Nanomaterials*.

Appenzeller, J. (2008). Carbon nanotubes for high-performance electronics—Progress and prospect. *Proceedings of the IEEE*, *96*(2), 201–211. doi:10.1109/JPROC.2007.911051

Appenzeller, J., Lin, Y.-M., Knoch, J., Chen, Z., & Avouris, P. (2005). Comparing Carbon Nanotube Transistors—The Ideal Choice: A Novel Tunneling Device Design. *IEEE Transactions on Electron Devices*, *52*(12), 2568–2576. doi:10.1109/TED.2005.859654

Auth, Ch., & Plummer, J. D. (1997). Scaling theory for cylindrical, fully depleted, surrounding gate MOSFET's. *IEEE Electron Device Letters*, *18*(2), 74–76. doi:10.1109/55.553049

Avci, U. E., Rios, R., Kuhn, K., & Young, I. A. (2011). Comparison of performance, switching energy and process variations for the TFET and MOSFET in logic. *2011 Symposium on VLSI Technology-Digest of Technical Papers*, 124-125.

Avouris, P., Radosavljević, M., & Wind, S. J. (2005). Carbon nanotube electronics and optoelectronics. In *Applied Physics of Carbon Nanotubes* (pp. 227–251). Berlin: Springer. doi:10.1007/3-540-28075-8_9

Azhari, R., Shahhoseini, A., & Aghaei-liavali, F. (2015). Stability Improvement and consumption power of CNTFET-based ternary memory ell. *2015 IEEE International Conference on Research in Computational Intelligence and Communication Networks (ICICRCICN, 2015).*

Bachtold, P. (2001). Logic Circuits with Carbon Nanotube Transistors. *Science*, *294*(5545), 1317–1320. doi:10.1126cience.1065824 PMID:11588220

Bacon, R. (1960). Growth, structure, and properties of graphite whiskers. *Journal of Applied Physics*, *31*(2), 283–290. doi:10.1063/1.1735559

Bakoglu, H. B., & Meindl, J. D. (1985). Optimal interconnection circuits for VLSI. *IEEE Transactions on Electron Devices*, *32*(5), 903–909. doi:10.1109/T-ED.1985.22046

Bala, S., & Khosla, M. (2018). Analysis of CNTFET for SRAM Cell Design. *Nanoscale Devices: Physics, Modeling, and Their Application*, 205.

Bala, S., & Khosla, M. (2018). Comparative Study and Analysis of CNTFET and Tunnel CNTFET. *Journal of Nanoelectronics and Optoelectronics*, *13*(3), 324–330. doi:10.1166/jno.2018.2234

Bala, S., & Khosla, M. (2018, September). Electrostatically doped tunnel CNTFET model for low-power VLSI circuit design. *Journal of Computational Electronics*, *5*, 1–8.

Balendhran, Walia, Nili, & Sriram. (2014). *Elemental analogues of graphene: Silicene, Germanene, Stanene and Phosphorene*. Wiley.

Balla, P. C., & Antoniou, A. (1984). Low power dissipation MOS ternary logic family. *IEEE Journal of Solid-State Circuits*, *19*(5), 739–749. doi:10.1109/JSSC.1984.1052216

Banerjee, K., & Srivastava, N. (2006, July). Are carbon nanotubes the future of VLSI interconnections? In *2006 43rd ACM/IEEE Design Automation Conference* (pp. 809-814). IEEE. 10.1145/1146909.1147116

Bao, M. (2005). *Analysis and design principles of MEMS devices*. Elsevier.

Barlian, A. A., Park, W. T., Mallon Jr, J. R., Rastegar, A. J., & Pruitt, B. L. (2009). Review: Semiconductor Piezoresistance for Microsystems 2009. *Proc. of the IEEE*, *97*(3)

Bastani, N. H., Moaiyeri, M. H., & Navi, K. (2018). An energy-and area-efficient approximate ternary adder based on CNTFET switching logic. *Circuits, Systems, and Signal Processing*, *37*(5), 1863–1883. doi:10.100700034-017-0627-1

Ben Jamaa, M. H., Mohanram, K., & De Micheli, G. (2011). An efficient gate library for ambipolar CNTFET logic. *IEEE Transactions on Computer-Aided Design of Integrated Circuits and Systems, 30*(2), 242–255. doi:10.1109/TCAD.2010.2085250

Berkeley Predictive Technology Model (BPTM). (n.d.). Available http://wwwdevice.eecs.berkeley.edu/˜ptm/

Bhat, K. N., & Nayak, M. M. (2013). MEMS pressure sensors-an overview of challenges in technology and packaging. *Journal of Smart Structures and Systems, 2*, 1-10.

Bhat, N. S. (2012). Design and modelling of different SRAM's based on CNTFET 32nm technology. *International Journal of VLSI Design & Communication Systems 3*(12).

Bhatnagar, S., Merkley, A., Berdine, R., Li, Y., & Roy, S. (2018, December). Variability-Aware Performance Assessment of Multi-Walled Carbon Nanotube Interconnects using a Predictor-Corrector Polynomial Chaos Scheme. In *2018 IEEE Electrical Design of Advanced Packaging and Systems Symposium (EDAPS)* (pp. 1-3). IEEE. 10.1109/EDAPS.2018.8680890

Bonaccorso, F. (2010). Graphene photonics and optoelectronics. *Nature Photonics*. doi:10.1038/nphoton.2010.186

Boucart, K., & Ionescu, A. M. (2007). Length scaling of the double gate tunnel FET with a high-k gate dielectric. *Solid-State Electronics, 51*(11), 1500–1507. doi:10.1016/j.sse.2007.09.014

Bourlon, B., Miko, C., Forro, L., Glattli, D. C., & Bachtold, A. (2004). Determination of the intershell conductance in multiwalled carbon nanotubes. *Physical Review Letters, 93*(17), 176806. doi:10.1103/PhysRevLett.93.176806 PMID:15525107

Burke, P. J. (2002). Luttinger liquid theory as a model of the gigahertz electrical properties of carbon nanotubes. *IEEE Transactions on Nanotechnology, 1*(3), 129–144. doi:10.1109/TNANO.2002.806823

Cakici, T., Kim, K., & Roy, K. (2007). FinFET Based SRAM Design for Low Standby Power Applications. *8th International Symposium on Quality Electronic Design*, 127-132. 10.1109/ISQED.2007.76

Chandrakasan, P., Sheng, S., & Brodersen, R. W. (1992). Low-Power CMOS Digital Design. *IEEE Journal of Solid-State Circuits, 27*(4), 473–484. doi:10.1109/4.126534

Chen, D., Chilstedt, S., Dong, C., & Pop, E. (2010). *What Everyone Needs to Know about Carbon-Based Nano circuits*. DAC.

Cheng. (2004). The impact of random doping effects on CMOS SRAM cell. *Proc. ESSCIRC*, 219–222.

Chen, L., Feng, B., & Wu, K. (2013). Observation of a possible superconducting gap in silicene on Ag(111) surface. *Applied Physics Letters, 102*(8), 081602. doi:10.1063/1.4793998

Cheung, C. L., Kurtz, A., Park, H., & Lieber, C. M. (2002, February). Diameter-controlled synthesis of carbon nanotubes. *The Journal of Physical Chemistry B*, *106*(10), 2429–2433. doi:10.1021/jp0142278

Choi, W. B., Chung, D. S., Kang, J. H., Kim, H. Y., Jin, Y. W., Han, I. T., ... Kim, J. M. (1999). Fully sealed, high-brightness carbon-nanotube field-emission display. *Applied Physics Letters*, *75*(20), 3129–3131. doi:10.1063/1.125253

Choi, W., & Park, J. (2016). A charge-recycling assist technique for reliable and low power SRAM Design. *IEEE Transactions on Circuits and Systems*, *63*(8), 1164–1175. doi:10.1109/TCSI.2016.2589118

Chowdhury, A. H., Akhter, N., & Al Faisal, A. (2012). Performance Analysis and Development of Self-Consistent Model of Carbon Nanotube Field Effect Transistor (CNTFET). In *Proc. of the Global Engg, Sci. and Tech. conf* (pp. 28-29). Academic Press.

Clifford, J. P., John, D. L., Castro, L. C., & Pulfrey, D. L. (2004). Electrostatics of partially gated carbon nanotube FETs. *IEEE Transactions on Nanotechnology*, *3*(2), 281–286. doi:10.1109/TNANO.2004.828539

Colinge, J. P. (2008). The new generation of SOI MOSFETs. *Romanian Journal of Information Science and Technology*, *11*(1), 3–15.

Das, D., & Rahaman, H. (2012, July). Unified model for analyzing timing delay and crosstalk effects in carbon nanotube interconnects. In *2012 4th Asia Symposium on Quality Electronic Design (ASQED)* (pp. 100-109). IEEE. 10.1109/ACQED.2012.6320484

Das, P. K., Majumder, M. K., & Kaushik, B. K. (2014). Dynamic crosstalk analysis of mixed multi-walled carbon nanotube bundle interconnects. *The Journal of Engineering*, *2014*(5), 227-233.

Dávila, M. E., & Le Lay, G. (2016). Few layer epitaxial germanene: A novel two-dimensional material Dirac material. *Scientific Reports*, *6*(1), 20714. doi:10.1038rep20714 PMID:26860590

de Alba Rosano, M. R., & Garcia-Garcia, A. D. (2006). Measuring leakage power in nanometer CMOS 6T SRAM cells. *Proc. IEEE Int. Conf. Reconfigurable, Comput. FPGA's.*

Deng, J., & Wong, H. S. P. (2007). A compact SPICE model for carbon-nanotube field-effect transistors including nonidealities and its application - Part I: Model of the intrinsic channel region. *IEEE Transactions on Electron Devices*, *54*(12), 3186–3194. doi:10.1109/TED.2007.909030

Deng, J., & Wong, H. S. P. (2007). A compact SPICE model for carbon-nanotube field-effect transistors including nonidealities and its application- Part II: Full device model and circuit performance benchmarking. *IEEE Transactions on Electron Devices*, *54*(12), 3195–3205. doi:10.1109/TED.2007.909043

Deng, J., & Wong, H.-S. P. (2006). A Circuit-Compatible SPICE model for Enhancement Mode Carbon Nanotube Field Effect Transistors. *Proc. Intl. Conf. Simulation of Semiconductor Processes and Devices*, 166 – 169. 10.1109/SISPAD.2006.282864

Dresselhaus, M. S., Chen, G., Tang, M. Y., Yang, R. G., Lee, H., Wang, D. Z., ... Gogna, P. (2007). New directions for low-dimensional thermoelectric materials. *Advanced Materials, 19*(8), 1043–1053. doi:10.1002/adma.200600527

Dresselhaus, M. S., Dresselhaus, G., & Saito, R. (1995). Physics of carbon nanotubes. *Carbon, 33*(7), 883–891. doi:10.1016/0008-6223(95)00017-8

Fiori, G., & Iannaccone, G. (2007). Simulation of Graphene Nanoribbon Field-Effect Transistors. *IEEE Electron Device Letters*. doi:10.1109/LED.2007.901680

Franklin, A. D., Luisier, M., Han, S.-J., Tulevski, G., Breslin, C. M., Gignac, L., ... Haensch, W. (2012). Sub-10 nm Carbon Nanotube Transistor. *Nano Letters, 12*(2), 758–762. doi:10.1021/nl203701g PMID:22260387

Gaikwad, V. T., & Deshmukh, P. R. (2015, January). Design of CMOS ternary logic family based on single supply voltage. In *2015 International Conference on Pervasive Computing (ICPC)* (pp. 1-6). IEEE. 10.1109/PERVASIVE.2015.7087114

Gholipour, M., & Masoumi, N. (2012). Efficient inclusive analytical model for delay estimation of multi-walled carbon nanotube interconnects. *IET Circuits, Devices & Systems, 6*(4), 252–259. doi:10.1049/iet-cds.2011.0283

Giulianini & Motta. (1970). *Polymer Self-assembly on Carbon Nanotubes*. doi:10.1007/978-1-4614-0742-3_1

Giustiniani, A., Tucci, V., & Zamboni, W. (2010, August). Modeling issues and performance analysis of high-speed interconnects based on a bundle of SWCNT. *IEEE Transactions on Electron Devices, 57*(8), 1978–1986. doi:10.1109/TED.2010.2050836

Gopalakrishnan, K., Griffin, P. B., & Plummer, J. D. (2002). I-MOS: a novel semiconductor device with a subthreshold slope lower than kT/q. *International Electron Devices Meeting*, 289-292. 10.1109/IEDM.2002.1175835

Gopalakrishnan, K., Woo, R., Jungemann, C., Griffin, P. B., & Plummer, J. D. (2005). "Impact ionization MOS (I-MOS)" Part II: Experimental results. *IEEE Transactions on Electron Devices, 52*(1), 77–84. doi:10.1109/TED.2004.841344

Guo, J., Datta, S., Lundstrom, M., Brink, M., McEuen, P., Javey, A., . . . McIntyre, P. (2002). Assessment of silicon MOS and carbon nanotube FET performance limits using a general theory of ballistic transistors. *Digest International Electron Devices Meeting, IEEE*, 711-714.

Guzmán-Verr, G. G., & Lew Yan Voon, L. C. (2007). Electronic structure of silicon-based nanostructures. *Physical Review B: Condensed Matter and Materials Physics, 76*(7), 075131. doi:10.1103/PhysRevB.76.075131

Haruehanroengra, S., & Wang, W. (2007, August). Analyzing conductance of mixed carbon-nanotube bundles for interconnect applications. *IEEE Transactions on Electronic Devices Letters, 28*(8), 756-759.

Hashempour, H., & Lombardi, F. (2008). Device model for ballistic CNFETs using the first conducting band. *IEEE Design & Test of Computers*, *25*(2), 178–186. doi:10.1109/MDT.2008.34

Hazeghi, A., Krishnamohan, T., & Philip Wong, H.-S. (2007). Schottky-barrier carbon nanotube field-effect transistor modeling. *IEEE Transactions on Electron Devices*, *54*(3), 439–445. doi:10.1109/TED.2006.890384

Heung, A., & Mouftah, H. T. (1985). Depletion/enhancement CMOS for a lower power family of three-valued logic circuits. *IEEE Journal of Solid-State Circuits*, *20*(2), 609–616. doi:10.1109/JSSC.1985.1052354

Hien, D. S., Luong, N. T., Thi, T. A. T., & Nga, D. V. (2009). 3D simulation of coaxial carbon nanotube field effect transistor. *Journal of Physics: Conference Series*, *187*, 012061. doi:10.1088/1742-6596/187/1/012061

Hirlekar, Yamagar, Garse, Vij, & Kadam. (2009). Carbon nanotubes and its applications: a review. *Asian Journal of Pharmaceutical and Clinical Research*, *2*(4), 17-27.

Hoenlein, W., Kreupl, F., Duesberg, G. S., Graham, A. P., Liebau, M., Seidel, R., & Unger, E. (2003). Carbon nanotubes for microelectronics: Status and future prospects. *Materials Science and Engineering C*, *23*(6-8), 663–669. doi:10.1016/j.msec.2003.09.153

Hone, J., Whitney, M., Piskoti, C., & Zettl, A. (1999, January). Thermal conductivity of single-walled carbon nanotubes. *Physical Review B: Condensed Matter*, *59*(4), R2514–R2516. doi:10.1103/PhysRevB.59.R2514

Hosseini, A., & Shabro, V. (2010). Thermally-aware modeling and performance evaluation for single-walled carbon nanotube-based interconnects for future high performance integrated circuits. *Microelectronic Engineering*, *87*(10), 1955–1962. doi:10.1016/j.mee.2009.12.008

Hsu, T. R. (2008). *MEMS and microsystems: design, manufacture, and nanoscale engineering*. John Wiley & Sons.

Huang, B. (2007). Making a field effect transistor on a single graphene nanoribbon by selective doping. *Applied Physics Letters*. doi:10.1063/1.2826547

Huang, W. Q., Li, Y., Akeb, H., & Li, C. (2004, October). Greedy algorithms for packing unequal circles into a rectangular container. *The Journal of the Operational Research Society*, *56*(5), 539–548. doi:10.1057/palgrave.jors.2601836

Hu, W., Xia, N., Wu, X., Liab, Z., & Yang, J. (2014). Silicene as a highly sensitive molecule sensor for NH3, NO and NO2. *Physical Chemistry Chemical Physics*.

Iijima, S. (1991). Helical microtubules of graphitic carbon. *Nature*, *354*(6348), 56–58. doi:10.1038/354056a0

Im, S., Srivastava, N., Banerjee, K., & Goodson, K. E. (2005). Scaling analysis of multilevel interconnect temperatures for high-performance ICs. *IEEE Transactions on Electron Devices*, *52*(12), 2710–2719. doi:10.1109/TED.2005.859612

International Technology Roadmap for semiconductor. (2013). Retrieved from http://public.itrs.net

ITRS. (2009). *Process Integration, Devices, and Structures*. Retrieved from http://www.itrs.net/Links/2009ITRS/2009Chapters_2009Tables/2009_PIDS.pdf

Jafarzadehpour, F., & Keshavarzian, P. (2016). Low-power consumption ternary full adder based on CNTFET. *IET Circuits, Devices & Systems*, *10*(5), 365–374. doi:10.1049/iet-cds.2015.0264

Jamaa, M. H. B., Mohanram, K., & Micheli, G. d. (2010). Power consumption of logic circuits in ambipolar carbon nanotube technology. In *Proceedings of Conference on Design, Automation & Test in Europe Conference & Exhibition* (pp. 303-306). IEEE.

Javey, Wang, Kim, & Dai. (2003). Advancements in complementary carbon nanotube field-effect transistor. *Proc. 2003 IEEE Int. Electron Devices Meeting*, 31.2.1–31.2.4. 10.1109/IEDM.2003.1269387

Javey, A., Guo, J., Wang, Q., Lundstrom, M., & Dai, H. (2003). Ballistic carbon nanotube field-effect transistors. *Nature*, *424*(6949), 654–657. doi:10.1038/nature01797 PMID:12904787

Javey, A., & Kong, J. (Eds.). (2009). *Carbon nanotube electronics*. Springer Science & Business Media.

Javey, A., Tu, R., Farmer, D., Guo, J., Gordon, R., & Dai, H. (2005). High performance n-type carbon nanotube field-effect transistors with chemically doped contacts. *Nano Letters*, *5*(2), 345–348. doi:10.1021/nl047931j PMID:15794623

Johnson, M., Somasekhar, D., Chiou, L.-Y., & Roy, K. (2002). Leakage control with efficient use of transistor stacks in single threshold CMOS. *IEEE Trans. VLSI Systems.*, *10*(1), 1–5. doi:10.1109/92.988724

Joshi, S., & Alabawi, U. (2017). Comparative Analysis of 6T, 7T, 8T, 9T, and 10T Realistic CNTFET Based SRAM. *Journal of Nanotechnology*, *2017*, 1–9. doi:10.1155/2017/4575013

Kang, Y., Kim, J., Kim, S., Shin, S., Jang, E. S., Jeong, J. W., . . . Kang, S. (2017, May). A novel ternary multiplier based on ternary CMOS compact model. In *2017 IEEE 47th International Symposium on Multiple-Valued Logic (ISMVL)* (pp. 25-30). IEEE. 10.1109/ISMVL.2017.52

Kang, S. M., & Leblebici, Y. (2002). *CMOS Digital Integrated Circuits Analysis and Design*. New York: McGraw-Hill.

Kaur, R., Singh, G., & Kaur, M. (2016). Effect of dielectric variations on performance of carbon nanotube field effect transistor based basic logic gates. *IOSR Journal of VLSI and Signal Processing*, *6*(4), 78–84. doi:10.9790/4200-0604027884

Kaushik, B. K., & Majumder, M. K. (2015). Carbon nanotube: Properties and applications. In *Carbon Nanotube Based VLSI Interconnects* (pp. 17–37). New Delhi: Springer.

Kaushik, B. K., Sarkar, S., Agarwal, R. P., & Joshi, R. C. (2010). An analytical approach to dynamic crosstalk in coupled interconnects. *Microelectronics Journal, 41*(2-3), 85–92. doi:10.1016/j.mejo.2009.12.011

Kaveh, A., & Mahdavi, V. R. (2014, May). Colliding bodies optimization: A novel meta-heuristic method. *Computers & Structures, 139*(1), 18–27. doi:10.1016/j.compstruc.2014.04.005

Kaveh, A., & Mahdvi, V. R. (2014, January). Colliding bodies optimization method for optimum design of truss structures with continuous variables. *Advances in Engineering Software, 70*(1), 1–12. doi:10.1016/j.advengsoft.2014.01.002

Khandelwal, M., & Sharan, N. (2017). Design and Performance Comparison of CNTFET-Based Binary and Ternary Logic Inverter and Decoder With 32 nm CMOS Technology. In *Advances in Computer and Computational Sciences* (pp. 429–437). Singapore: Springer. doi:10.1007/978-981-10-3770-2_40

Kim, Nan, & Choi. (2010). Power Gating for Ultra-Low Voltage Nanometer ICs. *IEEE ISCAS.*

Kima, B., Leea, J., Namgunga, S., & Kima, J. (2012). DNA sensors based on CNT-FET with floating electrodes. *Sensors and Actuators. B, Chemical, 169*, 182–187. doi:10.1016/j.snb.2012.04.063

Kim, W., Javey, A., Tu, R., Cao, J., Wang, Q., & Dai, H. (2005). Electrical contacts to carbon nanotubes down to 1 nm in diameter. *Applied Physics Letters, 87*(17), 173101. doi:10.1063/1.2108127

Kim, Y. B. (2010). Challenges for nanoscale MOSFETs and emerging nanoelectronics. *Transsanctions Electronic Material, 11*(3), 93–105. doi:10.4313/TEEM.2010.11.3.093

Kleene, S. C. (1954). *Introduction to metamathematics*. Academic Press.

Knoch, J., & Appenzeller, J. (2008). Tunneling phenomena in carbon nanotube field-effect transistors. *Physica Status Solidi. A, Applications and Materials Science, 205*(4), 679–694. doi:10.1002/pssa.200723528

Koo, K. H., Cho, H., Kumar, P., & Saraswat, K. C. (2007, October). Performance comparisons between carbon nanotubes, optical, and Cu for future high-performance on-chip interconnect applications. *IEEE Transactions on Electron Devices, 54*(12), 3206–3215. doi:10.1109/TED.2007.909045

Koswatta, S. O., Nikonov, D. E., & Lundstrom, M. S. (2005). Computational study of carbon nanotube p–i–n Tunnel FET. *IEEE Internat Electron Devices Meeting, Technical Digest*, 518-521. 10.1109/IEDM.2005.1609396

Kumar, S. S., Ojha, A. K., Nambisan, R., Sharma, A. K., & Pant, B. D. (2013). Design and simulation of MEMS silicon piezoresistive pressure sensor for barometric applications. *Proceedings of the ARTCom&ARTEEPEIE&itSIP and PCIE*, 339-345.

Kumar, H., Srivastava, S., & Singh, B. (2019). Comparative Analysis of 6T, 7T Conventional CMOS and CNTFET Based SRAM Cell Design. *Advanced Science Engineering in Medicine, 11*(1-2), 3–10.

Kumari, A., Rani, S., & Singh, B. (2018, July). I-V Characteristics analysis of Carbon Nanotube Field effect transistor and Graphene Nanotube Field-effect transistor. *IEEE Conference KIIT Bhubaneswar.*

Kumar, J., Majumder, M. K., Kaushik, B. K., Dasgupta, S., & Manhas, S. K. (2012, December). Novel modeling approach for multi-walled CNT bundle in global VLSI interconnects. In *2012 International Conference on Communications, Devices and Intelligent Systems (CODIS)* (pp. 476-479). IEEE. 10.1109/CODIS.2012.6422242

Kumar, M. G., Agrawal, Y., & Chandel, R. (2016, April). Carbon nanotube interconnects- a promising solution for VLSI circuits. *IETE Journal of Education*, *57*(2), 46–64. doi:10.1080/09747338.2016.1158129

Kumar, M. G., Chandel, R., & Agrawal, Y. (2017). An efficient crosstalk model for coupled multiwalled carbon nanotube interconnects. *IEEE Transactions on Electromagnetic Compatibility*, *60*(2), 487–496. doi:10.1109/TEMC.2017.2719052

Kumar, S. S., & Pant, B. D. (2015). Polysilicon thin film piezoresistive pressure microsensor: Design, fabrication and characterization. *Microsystem Technologies*, *21*(9), 1949–1958. doi:10.100700542-014-2318-1

Kumar, V. R., Kaushik, B. K., & Patnaik, A. (2015, December). Improved crosstalk noise modeling of MWCNT interconnects using FDTD technique. *Microelectronics Journal*, *46*(12), 1263–1268. doi:10.1016/j.mejo.2015.10.009

Kuresh & Hasan. (2009). Performance comparison of CNFET- and CMOS-based 6T SRAM cell in deep submicron. *Microelectronics Journal, 40*(6), 979-982.

Lee, S. Y., Kim, S., & Kang, S. (2019, May). Ternary Logic Synthesis with Modified Quine-McCluskey Algorithm. In *2019 IEEE 49th International Symposium on Multiple-Valued Logic (ISMVL)* (pp. 158-163). IEEE. 10.1109/ISMVL.2019.00035

Lee, C. S., Pop, E., Franklin, V. W., Haensch, W., & Wong, H.-S. P. (2015). A compact virtual source model for carbon nanotube FETs in the sub-10-nm regime—Part I: Intrinsic element. *IEEE Transactions on Electron Devices*, *62*(9), 3061–3069. doi:10.1109/TED.2015.2457453

Lee, J. U., Gipp, P. P., & Heller, C. M. (2004). Carbon nanotube p-n junction diodes. *Applied Physics Letters*, *85*(1), 145–147. doi:10.1063/1.1769595

Leonard, F., & Tersoff, J. (2000). Role of Fermi-level pinning in nanotube Schottky diodes. *Physical Review Letters*, *84*(20), 4693–4696. doi:10.1103/PhysRevLett.84.4693 PMID:10990773

Liang, J., Chen, L., Han, J., & Lombardi, F. (2014). Design and evaluation of multiple valued logic gates using pseudo N-type carbon nanotube FETs. *IEEE Transactions on Nanotechnology*, *13*(4), 695–708. doi:10.1109/TNANO.2014.2316000

Li, H., Lu, W. G., Li, J. J., Bai, X. D., & Gu, C. Z. (2005). Multichannel ballistic transport in multiwall carbon nanotubes. *Physical Review Letters*, *95*(8), 086601. doi:10.1103/PhysRevLett.95.086601 PMID:16196881

Li, H., Xu, C., Srivastava, N., & Banerjee, K. (2009). Carbon nanomaterials for next-generation interconnects and passives: Physics, status, and prospects. *IEEE Transactions on Electron Devices*, *56*(9), 1799–1821. doi:10.1109/TED.2009.2026524

Li, H., Yin, W.-Y., Banerjee, K., & Mao, J.-F. (2008). Circuit modeling and performance analysis of multi-walled carbon nanotube interconnects. *IEEE Transactions on Electron Devices*, *55*(6), 1328–1337. doi:10.1109/TED.2008.922855

Lin, Kim, & Lombardi. (2008). A New SRAM Cell Design Using CNTFETs. *International SoC Design Conference, 2008, 1*, I-168 - I-171.

Lin, S., Kim, Y. B., & Lombardi, F. (2011). CNTFET-based design of ternary logic gates and arithmetic circuits. *IEEE Transactions on Nanotechnology*, *10*(2), 217–225. doi:10.1109/TNANO.2009.2036845

Lin, S., Kim, Y. B., & Lombardi, F. (2012). Design of a ternary memory cell using CNTFETs. *IEEE Transactions on Nanotechnology*, *11*(5), 1019–1025. doi:10.1109/TNANO.2012.2211614

Lin, Y.-M., Appenzeller, J., Knoch, J., & Avouris, P. (2005). High performance carbon nanotube field-effect transistor with tunable polarities. *IEEE Transactions on Nanotechnology*, *4*(5), 481–489. doi:10.1109/TNANO.2005.851427

Litoria, P., Sandha, K. S., & Kansal, A. (2017). Impact of tunneling conductance on the performance of multi walled carbon nanotubes as VLSI interconnects for nano-scaled technology nodes. *Journal of Materials Science Materials in Electronics*, *28*(6), 4818–4827. doi:10.100710854-016-6128-3

Liu, C.-C., Feng, W., & Yao, Y. (2011). *Quantum Spin Hall Effect in Silicene and Two-Dimensional Germanium*. American Physical Society.

Li, X., Wang, X., Zhang, L., Lee, S., & Dai, H. (2008). Chemically derived, ultrasmooth graphene nanoribbon semiconductors. *Science*, *319*(5867), 1229–1232. doi:10.1126cience.1150878 PMID:18218865

Loumi, A., Ghammaz, A., & Belahrach, H. (2019, April). Crosstalk and delay analysis in Multi-Walled Carbon Nanotube and Single-Walled Carbon Nanotube Interconnects. In *2019 International Conference on Wireless Technologies, Embedded and Intelligent Systems (WITS)* (pp. 1-4). IEEE. 10.1109/WITS.2019.8723814

Łukasiewicz, J. (1970). *Selected Works*. *L.* Borkowski.

Mahar, B., Laslau, C., Yip, R., & Sun, Y. (2007). Development of carbon nanotube-based sensors—A review. *IEEE Sensors Journal*, *7*(2), 266–284. doi:10.1109/JSEN.2006.886863

Majumder, M. K., Das, P. K., Kaushik, B. K., & Manhas, S. K. (2012, December). Optimized delay and power performances for multi-walled CNT in global VLSI interconnects. In *2012 5th International Conference on Computers and Devices for Communication (CODEC)* (pp. 1-4). IEEE. 10.1109/CODEC.2012.6509358

Majumder, M. K., Kaushik, B. K., & Manhas, S. K. (2011, September). Comparison of propagation delay characteristics for single-walled CNT bundle and multiwalled CNT in global VLSI interconnects. In 2011 IEEE Recent Advances in Intelligent Computational Systems (pp. 911-916). IEEE.

Majumder, M.K., Das, P.K., Kumar, V.R., & Kaushik, B.K. (2015, August). Crosstalk induced delay analysis of randomly distributed mixed CNT bundle interconnect. *Journal of Circuits, Systems and Computers, 24*(10), 1550145-1- 1550145-15.

Majumder, M. K., Kaushik, B. K., & Manhas, S. K. (2014). Analysis of delay and dynamic crosstalk in bundled carbon nanotube interconnects. *IEEE Transactions on Electromagnetic Compatibility*, *56*(6), 1666–1673. doi:10.1109/TEMC.2014.2318017

Majumder, M. K., Kaushik, B. K., Manhas, S. K., & Das, P. K. (2012, December). Dynamic crosstalk effect in CNT bus architecture. In *2012 International Conference on Emerging Electronics* (pp. 1-4). IEEE. 10.1109/ICEmElec.2012.6636252

Majumder, M. K., Kumar, J., Kumar, V. R., & Kaushik, B. K. (2014, October). Performance analysis for randomly distributed mixed carbon nanotube bundle interconnects. *Micro & Nano Letters*, *9*(11), 792–796. doi:10.1049/mnl.2014.0502

Majumder, M. K., Pandya, N. D., Kaushik, B. K., & Manhas, S. K. (2012). Analysis of MWCNT and bundled SWCNT interconnects: Impact on crosstalk and area. *IEEE Electron Device Letters*, *33*(8), 1180–1182. doi:10.1109/LED.2012.2200872

Majumder, M. K., Pandya, N. D., Kaushik, B. K., & Manhas, S. K. (2012). Comparison of crosstalk delay between single and bundled SWNT for global VLSI interconnects. In *1st International Conference on Recent Advances in Information Technology (RAIT)*, (pp. 564-569), IEEE. 10.1109/RAIT.2012.6194590

Majumder, M. K., Pandya, N. D., Kaushik, B. K., & Manhas, S. K. (2012). Dynamic crosstalk effect in mixed CNT bundle interconnects. *Electronics Letters*, *48*(7), 384–385. doi:10.1049/el.2012.0536

Majumder, M. K., Pandya, N. D., Kaushik, B. K., & Manhas, S. K. (2012, March). Analysis of crosstalk delay and area for MWNT and bundled SWNT in global VLSI interconnects. In *Thirteenth International Symposium on Quality Electronic Design (ISQED)* (pp. 291-297). IEEE. 10.1109/ISQED.2012.6187508

Marani, R., & Perri, A. G. (2019). A Comparison of CNTFET and CMOS Technology through the Design of a SRAM Cell. *ECS Journal of Solid State Science and Technology*, *8*(1), M1–M18. doi:10.1149/2.0141901jss

Mas-Balleste, R., Gomez-Navarro, C., Gomez-Herrero, J., & Zamora, F. (2011). 2D materials: To graphene and beyond. *Nanoscale*, *3*(1), 20–30. doi:10.1039/C0NR00323A PMID:20844797

McEuen, P. L., Fuhrer, M. S., & Park, H. (2002). Single-walled carbon nanotube electronics. *IEEE Transactions on Nanotechnology*, *99*(1), 78–85. doi:10.1109/TNANO.2002.1005429

Meindl, J. D. (1995). Low power Microelectronics: Retrospect and Prospect. *Proceedings of the IEEE, 83*(4), 619–635. doi:10.1109/5.371970

Mittal, J., & Lin, K. L. (2017). Carbon nanotube-based interconnections. *Journal of Materials Science, 52*(2), 643–662. doi:10.100710853-016-0416-4

Moaiyeri, M. H., Nasiri, M., & Khastoo, N. (2016). An efficient ternary serial adder based on carbon nanotube FETs. *Engineering Science and Technology, an International Journal, 19*(1), 271-278.

Moaiyeri, M. H., Doostaregan, A., & Navi, K. (2011). Design of energy-efficient and robust ternary circuits for nanotechnology. *IET Circuits, Devices & Systems, 5*(4), 285–296. doi:10.1049/iet-cds.2010.0340

Moaiyeri, M. H., Mirzaee, R. F., Doostaregan, A., Navi, K., & Hashemipour, O. (2013). A universal method for designing low-power carbon nanotube FET-based multiple-valued logic circuits. *IET Computers & Digital Techniques, 7*(4), 167–181. doi:10.1049/iet-cdt.2013.0023

Mohamad, M., Soin, N., & Ibrahim, F. (2016, December). Design of a high sensitivity MEMS piezoresistive intracranial pressure sensor using three turns meander shaped piezoresistors. In *2016 International conference on bio-engineering for smart technologies (BioSMART)* (pp. 1-4). IEEE. 10.1109/BIOSMART.2016.7835596

Moradinasab, M., & Fathipour, M. (2009). Stable, low power and high performance SRAM based on CNFET. *10th International Conference on Ultimate Integration of Silicon (ULIS 2009)*, 317 – 320.

Morris Mano, M., & Ciletti, M. D. (2012). *Digital Design: With an Introduction to the Verilog HDL*. Academic Press.

Mukaidono, M. (1986). Regular ternary logic functions? ternary logic functions suitable for treating ambiguity. *IEEE Transactions on Computers, C-35*(2), 179–183. doi:10.1109/TC.1986.1676738

Murotiya, S. L., Gupta, A., & Pandit, A. (2014). CNTFET-based low power design of 4-input ternary XOR function. In *Proceedings of international conference on computer and communication technology* (pp. 347-350). IEEE.

Murotiya, S. L., Gupta, A., & Pandit, A. (2014). CNTFET-based low power design of 4-input ternary XOR function. *Proceedings of IEEE International Conference on Computer and Communication Technology (ICCCT)*, 347-350. 10.1109/ICCCT.2014.7001517

Naeemi, A., & Meindl, J. D. (2006). Compact physical models for multiwall carbon-nanotube interconnects. *IEEE Electron Device Letters, 27*(5), 338–340. doi:10.1109/LED.2006.873765

Naeemi, A., & Meindl, J. D. (2006). Design and performance modeling for single-walled carbon nanotubes as local, semiglobal, and global interconnects in gigascale integrated systems. *IEEE Transactions on Electron Devices, 54*(1), 26–37. doi:10.1109/TED.2006.887210

Naeemi, A., & Meindl, J. D. (2009). Carbon nanotube interconnects. *Annual Review of Materials Research*, *39*(1), 255–275. doi:10.1146/annurev-matsci-082908-145247

Nan, H., & Choi, K. (2010, November). Novel ternary logic design based on CNFET. In 2010 International SoC Design Conference (pp. 115-118). IEEE. doi:10.1109/SOCDC.2010.5682960

Nardelli, M., Fattebert, J.-L., Orlikowski, D., Roland, C., & Zhao, Q. (2000). Mechanical properties, defects, and electronic behavior of carbon nanotubes. Carbon, 38(11), 1703-1711.

Natsuki, T. (2017). Carbon nanotube-based nanomechanical sensor: Theoretical analysis of mechanical and vibrational properties. *Electronics (Basel)*, *6*(3), 56. doi:10.3390/electronics6030056

Navin, G., & Basavaraj, T. N. (2013). Factors affecting performance of CNT FET as a switch in memory cell. In *Proceedings of International Conference on Advanced Nano materials and Emerging Engineering Technologies* (pp. 655-657). IEEE. 10.1109/ICANMEET.2013.6609376

Novoselov, K. S., Geim, A. K., Morozov, S. V., Jiang, D., Zhang, Y., Dubonos, S. V., ... Firsov, A. A. (2004, October). Electric field effect in atomically thin carbon films. *Science*, *306*(5696), 666–669. doi:10.1126cience.1102896 PMID:15499015

Obitayo, W., & Liu, T. (2012). A review: Carbon nanotube-based piezoresistive strain sensors. *Journal of Sensors*.

Ostling, D., Tomanek, D., & Rosen, A. (1997). Electronic structure of single-wall, multiwall, and filled carbon nanotubes. *Physical Review B: Condensed Matter and Materials Physics*, *55*(20), 13980–13988. doi:10.1103/PhysRevB.55.13980

Ouyang, Yoon, & Guo. (2007). Scaling Behaviors of Graphene Nanoribbon FETs: A Three-Dimensional Quantum Simulation Study. *IEEE Transaction on Electronic Devices*.

Panda, A., & Pani, S. (2014, December). *A New model based on colliding bodies optimization for identification of Hammerstein Plant.* Paper presented at IEEE INDICON, Pune, India. 10.1109/INDICON.2014.7030381

Pandya, N. D., Majumder, M. K., Kaushik, B. K., & Manhas, S. K. (2012, May). Performance Comparision of Mixed CNT Bundle in Global VLSI Interconnect. In *2012 International Conference on Communication Systems and Network Technologies* (pp. 790-793). IEEE. 10.1109/CSNT.2012.171

Pant, P., De, V., & Chstterjee, A. (1997). Device-Circuit Optimization for Minimal Energy and Power Consumption in CMOS Random Logic Network. *ACM/IEEE Design Automation Conf.*, 403-408.

Patel, P.K., Malik, M.M., & Gupta, T.K. (2018). Reliable high-yield CNTFET-based 9T SRAM operating near threshold. *Journal of Computational Electronics, 17*(2), 774–783.

Patel, P. K., Malik, M. M., & Gupta, T. K. (2019). Performance evaluation of single-ended disturb-free CNTFET-based multi-Vt SRAM. *Microelectronics Journal*, *90*, 19–28. doi:10.1016/j.mejo.2019.05.015

Patel, R. S., Ganapathiraj, S., Kundaragi, S., & Gadekar, P. (2016). Carbon Nanotubes Based Piezoresistive Pressure Sensor. *Proceedings of 22nd IRF International Conference.*

Patil, N., Lin, A., Zhang, J., Wong, H.-S. P., & Mitra, S. (2009). Digital VLSI logic technology using carbon nanotube FETs: Frequently asked questions. *Proc. IEEE Design Autom. Conf.*, 304–309. 10.1145/1629911.1629995

Pop, E., Mann, D., Reifenberg, J., Goodson, K., & Dai, H. (2005, December). Electro-thermal transport in metallic single-wall carbon nanotubes for interconnect applications. In *IEEE InternationalElectron Devices Meeting, 2005. IEDM Technical Digest.* (pp. 4-pp). IEEE. Retrieved from www.ptm.asu.edu

Pop, E., Mann, D. A., Goodson, K. E., & Dai, H. (2007). Electrical and thermal transport in metallic single-wall carbon nanotubes on insulating substrates. *Journal of Applied Physics*, *101*(9), 093710. doi:10.1063/1.2717855

Pourfath, M., Kosina, H., & Selberherr, S. (2007). Tunneling CNTFETs. *Journal of Computational Electronics*, *6*(1-3), 243–246. doi:10.100710825-006-0099-1

Prabhu, S., & Sarwade, N. (2013). Application of CNTFET as logic gates and its implementation using HSPICE. *International Journal of Modern Engineering Research*, *3*, 3646–3648.

Predictive Technology Model. (n.d.). Retrieved from http://ptm.asu.edu/modelcard/LP/32nm_LP.pm

Pu, S. N., Yin, W. Y., Mao, J. F., & Liu, Q. H. (2009). Crosstalk prediction of single-and double-walled carbon-nanotube (SWCNT/DWCNT) bundle interconnects. *IEEE Transactions on Electron Devices*, *56*(4), 560–568. doi:10.1109/TED.2009.2014429

Qe, H., Kim, Y.-B., & Choi, M. (2012). *A High Speed Low Power Modulo 2^n+1 Multiplier Design Using Carbon-nanotube Technology. IEEE International Midwest Symposium on Circuits and Systems (MWCAS)*, Boise ID.

Qi, H., Kim, Y. B., & Choi, M. (2012). A high speed low power modulo 2n+1 multiplier design using carbon-nanotube technology. In *Proceedings of 55th International Midwest Symposium on Circuits and Systems* (pp. 406-409). IEEE.

Rahman, Guo, Datta, & Lundstrom. (2003). Theory of ballistic nanotransistors. *IEEE Trans. Electron Devices*, *50*(10), 1853–1864.

Rahman, A., Guo, J., Datta, S., & Lundstrom, M. S. (2003). Theory of ballistic nanotransistors. *IEEE Transactions on Electron Devices*, *50*(9), 1853–1864. doi:10.1109/TED.2003.815366

Raj, B., Saxena, A. K., & Dasgupta, S. (2009). Analytical modeling for the estimation of leakage current and subthreshold swing factor of nanoscale double gate FinFET device". *Microelectronics International*, *26*(1), 53–63. doi:10.1108/13565360910923188

Rani, S., Kaur, G., & Lakha, B. S. (2018). Ternary Logic Design Approach-From CMOS to CNTFET. *Journal of Nanoscience Nanoengineering and Applications*, *8*(2), 20–30.

Rani, S., & Singh, B. (2018). Investigation of Schottky barrier, conventional and tunnel carbon nanotube field effect transistor for low power design. *Journal of Nanoelectronics and Optoelectronics*, *13*(1), 76–82. doi:10.1166/jno.2018.2169

Rao, S. S. (1984). *Optimization: Theory and Applications*. New York: Wiley.

Raychowdhury, A., Mukhopadhyay, S., & Roy, K. (2004). A Circuit-Compatible Model of Ballistic Carbon Nanotube Field-Effect Transistors. *IEEE Transactions on Computer-Aided Design of Integrated Circuits and Systems*, *23*(10), 1411–1420. doi:10.1109/TCAD.2004.835135

Raychowdhury, A., & Roy, K. (2005). Carbon-nanotube-based voltage-mode multiple-valued logic design. *IEEE Transactions on Nanotechnology*, *4*(2), 168–179. doi:10.1109/TNANO.2004.842068

Romero. (2009). Adsorption of ammonia on graphene. *Nanotechnology*.

Rossi, D., Cazeaux, J. M., Metra, C., & Lombardi, F. (2007). Modeling crosstalk effects in CNT bus architectures. *IEEE Transactions on Nanotechnology*, *6*(2), 133–145. doi:10.1109/TNANO.2007.891814

Rossnagel, S. M., & Kuan, T. S. (2004). Alteration of Cu conductivity in the size effect regime. *Journal of Vacuum Science & Technology. B, Microelectronics and Nanometer Structures: Processing, Measurement, and Phenomena: An Official Journal of the American Vacuum Society*, *22*(1), 240–247.

Rutherglen, C., & Burke, P. (2009). Nanoelectromagnetics: Circuit and electromagnetic properties of carbon nanotubes. *Small, 5*(8), 884-906.

Sahoo, R., Sahoo, S. K., & Sankisa, K. C. (2015). Design of an efficient CNTFET using optimum number of CNT in channel region for logic gate implementation. In *Proceedings of International Conference on VLSI Systems, Architecture, Technology and Applications* (pp. 1-4). IEEE.

Sahoo, S. K., Akhilesh, G., Sahoo, R., & Muglikar, M. (2017). High-performance ternary adder using CNTFET. *IEEE Transactions on Nanotechnology*, *16*(3), 368–374. doi:10.1109/TNANO.2017.2649548

Samadi, H., Shahhoseini, A., & Aghaei-liavali, F. (2017). A new method on designing and simulating CNTFET_based ternary gates and arithmetic circuits. *Microelectronics Journal*, *63*, 41–48. doi:10.1016/j.mejo.2017.02.018

Sandha, K. S., & Sharma, S. (2018). Performance Analysis of Different Mixed-MWCNT Structures as VLSI Interconnects for Nano-Electronics IC Design. *Journal of Nanoelectronics and Optoelectronics*, *13*(3), 357–367. doi:10.1166/jno.2018.2229

Sankar, P. G., & Kumar, K. U. (2011). Mechanical and electrical properties of single walled carbon nanotubes: A computational study. *European Journal of Scientific Research*, *60*(3), 342–358.

Sarto, M. S., & Tamburrano, A. (2010, January). Single-conductor transmission line model of multiwall carbon nanotubes. *IEEE Transactions on Nanotechnology*, *9*(1), 82–92. doi:10.1109/TNANO.2009.2023516

Scarlet, S. P., Ambika, R., & Srinivasan, R. (2017). Effect of eccentricity on junction and junctionless based silicon nanowire and silicon nanotube FETs. *Superlattices and Microstructures*, *107*, 178–188. doi:10.1016/j.spmi.2017.04.015

Seabaugh, A. C., & Zhang, Q. (2010). Low-voltage tunnel transistors for beyond CMOS logic. *Proceedings of the IEEE*, *98*(12), 2095–2110. doi:10.1109/JPROC.2010.2070470

Selvin, S. A., Lovelin, S. A., Moorthy, N. B., Gupta, A., Alagappan, M., & Ramalingam, V. (2011). Design and Simulation of Carbon Nanotube based Piezoresistive Pressure sensor. In *Expert from the proceedings of the 2011 COMSOL Conference in Bangalore, Page* (No. 3). Academic Press.

Semiconductor Industry Association, International Technology Roadmap for Semiconductors (ITRS). (2012). Available: http://public.itrs.net

Sethi, D., Kaur, M., & Singh, G. (2017). Design and performance analysis of a CNFET-based TCAM cell with dual-chirality selection. *Journal of Computational Electronics*, *16*(1), 106–114. doi:10.100710825-017-0952-4

Shaker, A., Ossaimee, M., Zekry, A., & Abouelatta, M. (2015). Influenceof gate overlap engineering on ambipolar and high frequency characteristics of tunnel-CNTFET. *Superlattices and Microstructures*, *86*, 518–530. doi:10.1016/j.spmi.2015.08.008

Sharifi, F., Panahi, A., Moaiyeri, M. H., Sharifi, H., & Navi, K. (2018). High performance CNFET-based ternary full adders. *Journal of the Institution of Electronics and Telecommunication Engineers*, *64*(1), 108–115. doi:10.1080/03772063.2017.1338973

Sharma, S. K., Raj, B., & Khosla, M. (2017). Subthreshold Performance of $In_{1-x}Ga_xAs$ Based Dual Metal with Gate Stack Cylindrical/Surrounding Gate Nanowire MOSFET for Low Power Analog Applications. *Journal of Nanoelectronics and Optoelectronics.*, *12*(2), 171. doi:10.1166/jno.2017.1961

Shin, S., Jang, E., Jeong, J. W., Park, B. G., & Kim, K. R. (2015). Compact design of low power standard ternary inverter based on OFF-state current mechanism using nano-CMOS technology. *IEEE Transactions on Electron Devices*, *62*(8), 2396–2403. doi:10.1109/TED.2015.2445823

Shiraishi & Ata. (2000). Work function of carbon nanotubes. *MRS Online Proceedings Library Archive*, 633.

Shreya, S., & Chandel, R. (2014). Performance analysis of CNTFET based digital logic circuits. In *Proceedings of 2014 Students Conference on Engineering and Systems (SCES)*. Allahabad, India. IEEE. 10.1109/SCES.2014.6880063

Singh Duksh, Y., Kaushik, B. K., Sarkar, S., & Singh, R. (2013). Analysis of propagation delay and power with variation in driver size and number of shells in multi walled carbon nanotube interconnects. *Journal of Engineering, Design and Technology, 11*(1), 19-33.

Singh, A., Khosla, M., & Raj, B. (2016). Circuit compatible model for electrostatic doped Schottky barrier CNTFET. *Journal of Electronic Materials, 45*(10), 5381–5390. doi:10.100711664-016-4743-7

Singh, A., Khosla, M., & Raj, B. (2016). Comparative analysis of carbon nanotube field effect transistor and nanowire transistor for low power circuit design. *Journal of Nanoelectronics and Optoelectronics, 11*(3), 388–393. doi:10.1166/jno.2016.1913

Singh, A., Khosla, M., & Raj, B. (2017). Design and analysis of electrostatic doped Schottky barrier CNTFET based low power SRAM. *AEÜ. International Journal of Electronics and Communications, 80*, 67–72. doi:10.1016/j.aeue.2017.06.030

Singh, A., Saini, D. K., Agarwal, D., Aggarwal, S., Khosla, M., & Raj, B. (2016). Modeling and simulation of carbon nanotube field effect transistor and its circuit application. *Journal of Semiconductors, 37*(7), 074001. doi:10.1088/1674-4926/37/7/074001

Singh, J., Mohanty, S. P., & Pradhan, D. (2012). *Robust SRAM Designs and Analysis*. Springer Science & Business Media.

Singh, J., Ramakrishnan, K., Mookerjea, S., Datta, S., Vijaykrishnan, N., & Pradhan, D. (2010). A novel si-tunnel FET based SRAM design for ultra low-power 0.3V VDD applications. In *Proceedings of the 2010 Asia and South Pacific Design Automation Conference*. IEEE Press. 10.1109/ASPDAC.2010.5419897

Singh, K., & Raj, B. (2015). Influence of temperature on MWCNT bundle, SWCNT bundle and copper interconnects for nanoscaled technology nodes. *Journal of Materials Science Materials in Electronics, 26*(8), 6134–6142. doi:10.100710854-015-3193-y

Singh, K., & Raj, B. (2015). Performance and analysis of temperature dependent multi-walled carbon nanotubes as global interconnects at different technology nodes. *Journal of Computational Electronics, 14*(2), 469–476. doi:10.100710825-015-0667-3

Singh, K., & Raj, B. (2015). Temperature-dependent modeling and performance evaluation of multi-walled CNT and single-walled CNT as global interconnects. *Journal of Electronic Materials, 44*(12), 4825–4835. doi:10.100711664-015-4040-x

Singh, M. K., & Raj, B. (2015). Comparative analysis of carbon nanotube field effect transistors. *Proceedings of the IEEE 4th Global Conference on Consumer Electronics*, 552-555. 10.1109/GCCE.2015.7398601

Sinha, S. K., & Chaudhury, S. (2014). Comparative study of leakage power in CNTFET over MOSFET device. *Journal of Semiconductors, 35*(11), 114–02. doi:10.1088/1674-4926/35/11/114002

Sinha, S. K., & Chaudhury, S. (2015). Analysis of different parameters of channel material and temperature on threshold voltage of CNTFET. *Materials Science in Semiconductor Processing*, *31*, 431–438. doi:10.1016/j.mssp.2014.12.013

Smith, C. S. (1954). Piezoresistance effect in germanium and silicon. *Physical Review*, *94*(1), 42–49. doi:10.1103/PhysRev.94.42

Srinivasu, B., & Sridharan, K. (2016). Carbon nanotube FET-based low-delay and low-power multi-digit adder designs. *IET Circuits, Devices & Systems*, *11*(4), 352–364. doi:10.1049/iet-cds.2016.0013

Srivastava, N., Joshi, R. V., & Banerjee, K. (2005, December). Carbon nanotube interconnects: implications for performance, power dissipation and thermal management. In IEEE InternationalElectron Devices Meeting, 2005. IEDM Technical Digest. (pp. 249-252). IEEE. doi:10.1109/IEDM.2005.1609320

Srivastava, A., Xu, Y., & Sharma, A. K. (2010, April). Carbon nanotube for next generation very large scale integration interconnects. Journal of Nanophotonics. *Society of Photo-Optical Instrumentation Engineers*, *4*(1), 041690-1–041690-26.

Srivastava, N., & Banerjee, K. (2005, May). Performance analysis of carbon nanotube interconnects for VLSI applications. In *Proceedings of the 2005 IEEE/ACM International conference on Computer-aided design* (pp. 383-390). IEEE Computer Society. 10.1109/ICCAD.2005.1560098

Srivastava, N., Li, H., Kreupl, F., & Banerjee, K. (2009). On the applicability of single-walled carbon nanotubes as VLSI interconnects. *IEEE Transactions on Nanotechnology*, *8*(4), 542–559. doi:10.1109/TNANO.2009.2013945

Stanford University CNFET Model website. (n.d.). Available: http://nano.stanford.edu/model.php?id=23

Stanford University CNFET Model. (n.d.). Retrieved from http://nanostanford.edu/models.phpwebsite

Stanford University CNTFET Model Website. (2008). *Stanford University*. Available: http://nano.stanford.edu/model. php? id=23

Stanford University Nanoelectronics Group. (2018). *Stanford University CNFET model*. Retrieved from http://nano.stanford.edu/model. php?id=23

Steinhögl, W., Schindler, G., Steinlesberger, G., Traving, M., & Engelhardt, M. (2005). Comprehensive study of the resistivity of copper wires with lateral dimensions of 100 nm and smaller. *Journal of Applied Physics*, *97*(2), 023706. doi:10.1063/1.1834982

Suzuki, S., Bower, C., Watanabe, Y., & Zhou, O. (2000). Work functions and valence band states of pristine and Cs-intercalated single-walled carbon nanotube bundles. *Applied Physics Letters*, *76*(26), 4007–4009. doi:10.1063/1.126849

Synopsys for HSPICE tools. (2016). Available: http://www.synopsys.com

Sze, S. M. (2009). *Semiconductor Devices: Physics and Technology* (2nd ed.). Wiley India.

Tabrizchi, S., Azimi, N., & Keivan, N. (2016). Design a novel ternary half adder and multiplier based on carbon nano-tube field effect transistors (CNTFETs). *Frontiers*, *1*, 423–433.

Takeda, K., & Shiraishi, K. (1994). Theoretical possibility of stage corrugation in Si and Ge analogs of graphite. *Physical Review B: Condensed Matter and Materials Physics*. doi:10.1103/PhysRevB.50.14916

Tekleab, D., Tran, H. H., & Sleight, J. W. (2012). *Silicon nanotube MOSFET*. U.S. Patent 2012217468 A1.

Tekleab, D. (2014, May). Device performance of silicon nanotube field effect transistors. *IEEE Electron Device Letters*, *35*(5), 506–508. doi:10.1109/LED.2014.2310175

Tsai, Huang, Chang, Lin, Jeng, & Bansil. (2013). *Gated silicene as a tunable source of nearly 100% spin-polarized electrons*. NCBI.

Tuantranont, A., & Bright, V. M. (2000). Introduction to Micro-Electro-Mechanical Systems (MEMS) with Emphasis on Optical Applications. *NECTEC Technical Journal*, *1*(6), 227–236.

Uhlig, B., Liang, J., Lee, J., Ramos, R., Dhavamani, A., Nagy, N., . . . Asenov, A. (2018, March). Progress on carbon nanotube BEOL interconnects. In 2018 Design, Automation & Test in Europe Conference & Exhibition (DATE) (pp. 937-942). IEEE. doi:10.23919/DATE.2018.8342144

Uhlig, B., Dhavamani, A., Nagy, N., Lilienthal, K., Liske, R., Ramos, R., & Georgiev, V. (2018, June). Challenges and Progress on Carbon Nanotube Integration for BEOL Interconnects. In *2018 IEEE International Interconnect Technology Conference (IITC)* (pp. 16-18). IEEE. 10.1109/IITC.2018.8454842

Walia & Randhawa. (2018). Adsorption and dissociation of sulfur-based toxic gas molecules on silicene ASiNRs: a quest for high-performance gas sensors and catalysts. *Journal of Molecular Modelling*.

Walia, G. K., & Deep, K. K. R. (2014). First-principles investigation on defect-induced silicene ASiNRs — A superior media for sensing NH3, NO2 and NO gas molecules. In Surface Science. Elsevier.

Walia, G. K., & Deep, K. K. R. (2018). *Gas-sensing properties of Armchair silicene ASiNRs towards carbon-based gases with single-molecule resolution. In Structural Chemistry*. Springer.

Wang, W., & Choi, K. (2010). Novel curve fitting design method for carbon nanotube SRAM cell optimization. *IEEE International Conference on Electro/Information Technology (EIT)*.

Wang, W., Haruehanroengra, S., Shang, L., & Liu, M. (2007, April). Inductance of mixed carbon nanotube bundles. *Micro & Nano Letters*, *2*(2), 35–39. doi:10.1049/mnl:20070027

Wang, W., Yu, Z., & Choi, K. (2011). *High SNM 6T CNFET SRAM Cell Design Considering Nanotube Diameter and Transistor Ratio*. IEEE EIT. doi:10.1109/EIT.2011.5978618

Wang, X.-Q., Li, H.-D., & Wang, J.-T. (2012). Induced ferromagnetism in one-side semihydrogenated silicene and germanene. *Physical Chemistry Chemical Physics.*

Wang, Y., & Yeow, J. T. (2009). A review of carbon nanotubes-based gas sensors. *Journal of Sensors*, ▪▪▪, 2009.

Wei, Roy, & De. (2000). Low Voltage Low Power CMOS Design Techniques for Deep Submicron ICs. *13th International Conference on VLSI Design*, 24 – 29.

Wei, B. Q., Vajtai, R., & Ajayan, P. M. (2001). Reliability and current carrying capacity of carbon nanotubes. *Applied Physics Letters*, *79*(8), 1172–1174. doi:10.1063/1.1396632

Xu, C., Li, H., & Banerjee, K. (2009). Modeling, analysis and design of grapheme nanoribbon interconnects. *IEEE Transactions on Electron Devices*, *56*(8), 1567–1578. doi:10.1109/TED.2009.2024254

Yamauchi, H. (2007). Embedded SRAM Circuit Design Technologies for a 45 nm and Beyond. *The 7th International Conference on ASIC (ASICON).*

Yamauchi. (2008). Embedded SRAM Design and Trend. *IEEE ISSCC 2008, Memory forum Embedded Memory Design for Nano-scale VLSI system.*

Yang, X. (2010). Triple-mode single-transistor on a single graphene nanoribbon by selective doping. *Nano.*

Yeole, P., & Padole, D. V. (2015). Design of basic logic gates using carbon nano tube field effect transistor and calculation of figure of merit. In *Proceedings of 7th International Conference on Emerging Trends in Engineering & Technology* (pp. 28-33). IEEE. 10.1109/ICETET.2015.15

Yilmazoglu, O., Joshi, R., Popp, A., Pavlidis, D., & Schneider, J. J. (2011, January). Pronounced field emission from vertically aligned carbon nanotubes blocks and bundles. *Journal of Vacuum Science & Technology B Microelectronics and Nanometer Structures*, *29*(2), 02B106-1 02B106-5.

Young, K. K. (1989). Short-channel effect in fully depleted SOI mosfets. *IEEE Transactions on Electron Devices*, *36*(2), 399–402. doi:10.1109/16.19942

Yu, M. F., Lourie, O., Dyer, M. J., Moloni, K., Kelly, T. F., & Ruoff, R. S. (2000, January). Strength and breaking mechanism of multiwalled carbon nanotubes under tensile load. *Science*, *287*(5453), 637–640. doi:10.1126cience.287.5453.637 PMID:10649994

Yunus, N. A. M., Halin, I. A., Sulaiman, N., Ismail, N. F., & Sheng, O. K. (2015). Valuation on MEMS Pressure Sensors and Device Applications. *International Journal of Electrical, Computer, Energetic, Electronic and Communication Engineering*, *9*(8).

Yu, Z., Chen, Y., Nan, H., Wang, W., & Choi, K. (2011). *Design of Novel Low Power 6T CNFET SRAM Cell Working in Sub-Threshold Region.* IEEE EIT.

Zaminpayma, E., & Nayebi, P. (2016). Band gap engineering in silicene: A theoretical study of density functional tight-binding theory. *Physica E, Low-Dimensional Systems and Nanostructures*. doi:10.1016/j.physe.2016.06.016

Zare, Shadmani, & Pournamdari. (2013). DFT/NBO study of Nanotube and Calixarene with anti-cancer drug. *Journal of Nanostructure in Chemistry*.

Zhang, Y., Ichihashi, T., Landree, E., Nihey, F., & Iijima, S. (1999). Heterostructures of single-walled carbon nanotubes and carbide nanorods. *Science*, *285*(5434), 1719–1722. doi:10.1126cience.285.5434.1719 PMID:10481005

Zheng & Zhang. (2012). The electronic and magnetic properties of functionalized silicene: a first-principles study. *Nanoscale Res Lett.*, *7*(1), 422.

Zhu, L., Xu, J., Xiu, Y., Sun, Y., Hess, D. W., & Wong, C. P. (2006, February). Growth and electrical characterization of high-aspect-ratio carbon nanotube arrays. *Carbon*, *44*(2), 253–258. doi:10.1016/j.carbon.2005.07.037

About the Contributors

Balwinder Raj was born in Pathankot, a city of Punjab, India, on 25 Nov. 1980. He did B. Tech, Electronics Engineering (PTU Jalandhar), M. Tech-Microelectronics (PU Chandigarh) and Ph.D-VLSI Design (IIT Roorkee), India in 2004, 2006 and 2010 respectively. For further research work, European Commission awarded him "Erasmus Mundus" Mobility of life research fellowship for postdoc research work at University of Rome, Tor Vergata, Italy in 2010-2011. Dr. Raj also received India4EU (India for European Commission) Fellowship and worked as visiting researcher at KTH University, Stockholm, Sweden, Oct-Nov 2013. Currently, he is working as Assistant Professor, Department of ECE, NITTTR Chandigarh. I had worked at National Institute of Technology (NIT Jalandhar) (An Institute of national Importance, MHRD), Punjab, India from 1 May 2012 to 16 December 2019. Dr. Raj worked as Assistant Professor at ABV-IIITM Gwalior (An autonomous institute established by Ministry of HRD, Govt. of India) July 2011 to Apl 2012. Dr. Raj has authored/co-authored two books, three book chapters and more than 60 research papers in peer reviewed international/national journals and conferences. His areas of interest in research are Nanoscale Semiconductor Device Modeling

Mamta Khosla is presently working as Associate Professor in the Department of Electronics and Communication Engineering, National Institute of Technology, Jalandhar. India. Her areas of interest are nanoscale Devices, digital design, soft computing and realization of intelligent systems. She received her Bachelor of Technology from National Institute of Technology, Kurukshetra, Haryana (India) with specialization in Electronics & Communication Engineering and Master of Technology from Punjab Technical University, Jalandhar with distinction in Electronics & Communication Engineering. Her name has been inscribed in Academic Roll of Honour of Guru Harkrishan Educational Society, Chandigarh for securing first position in Punjab Technical University, Jalandhar in Master of Technology.

Amandeep Singh received Ph.D-VLSI Design from NIT Jalandhar, India in 2017. Currently, he is working with the Department of Electronics and Communication as Assistant Professor at National Institute of Technology Srinagar, J & K, India. His research interest includes Semiconductor Device Modeling, Low Power VLSI Design, Reconfigurable Computing and its FPGA Implementation.

* * *

Yash Agrawal is faculty at Dhirubhai Ambani Institute of Information and Communication Technology (DA-IICT) Gandhinagar, Gujarat, India since 2016. He has received his Ph.D. and M.Tech. degrees in VLSI specialization of E&CE department of NIT Hamirpur in 2016 and 2012 respectively. He has graduated his bachelor's in Engineering from KITS Ramtek, Maharashtra in 2009. His research interest includes digital VLSI, nanotechnology, analytical modeling, simulation and analysis of VLSI circuits and on-chip interconnects. In the last five years, he has been dedicatedly working on graphene FETs, graphene interconnects, high performance computing systems, optimization schemes for mixed CNTs, numerical method based FDTD modeling for VLSI circuits, development of GaAs based devices and systems, image processing using VLSI and FPGA. He has several achievements to his credits. To name a few, he has been awarded with prestigious IETE—K S Krishnan Memorial Award-2017 for the Best System Oriented research paper. He achieved the third place in all India Mentor Graphics design contest held at Bangalore, India, in 2011. He has several publications in Book Chapters of Springer, Journals including IEEE Transactions in — Electromagnetic Compatibility, —Nanotechnology, Springer, Taylor and Francis and several national and international reputed Conferences. He is the Editorial board member and Reviewer of various journals of repute.

Tarun Attri Student, Electronics and Communication engineering department Guru Nanak Dev University Regional Campus Jalandhar. Area of interest- Nano-electronics and optical communication.

Rajeevan Chandel was awarded Ph.D. degree from Indian Institute of Technology (IIT) Roorkee, India, under QIP scheme of Govt. of India in 2005. She received M.Tech. degree in Integrated Electronics and Circuits from IIT Delhi, India, in 1997. She received B.E. degree in E&CE from TIET (Thapar University) Patiala, India, in 1990. She is a double gold medalist of Himachal Pradesh University, Shimla, India, in Pre-University and Pre-Engineering in 1985 and 1986, respectively. She joined the Department of Electronics and Communication Engineering, NIT Hamirpur, HP India, as a lecturer in 1990, where presently she is working as a professor and has been the Head of the Department twice. She is presently Dean (Research &

Consultancy). She has six sponsored research projects to her credit and over fifty research publications in international Journals of Repute. Her research interests include electronic circuit modelling and low-power VLSI design. Dr. Chandel is a Fellow of IETE, Member IEEE and Life Member of ISTE.

Rekha Devi is an Assistant professor in Chandigarh university Mohali in Electronics & communication engineering and pursuing a Ph. D from I.K.G.P.T.U. She has more than 8 years of teaching experience.

Ashwani Kumar received his Ph.D. degree in Electrical Engineering from Indian Institute of Technology (IIT) Roorkee, India in 2005. He received M.E. degree in Power Systems from PEC, Chandigarh and B.Tech. in Electrical Engineering from Kerala University, India. Dr. Kumar joined the Department of Electrical Engineering, NIT Hamirpur, HP India, as Lecturer in 1991, where presently he is working as a Professor. He has been Head of Electrical Engg. Department and is presently Coordinator QIP & IE(I) at NIT Hamirpur. He has four sponsored research projects to his credit and over one hundred and fifty research publications in various international Journals of Repute and Conferences. His research areas are harmonic estimation and elimination, condition monitoring of transformers. He is a Fellow of IETE, Member IEEE and Life Member of ISTE.

Raj Kumar is pursuing his Ph.D. from Panjab University, Chandigarh. he has done Mtech in VLSI Design from CDAC, Mohali, INDIA. His research interests are the design and modeling of advanced multigate 3D devices.

Yadwinder Kumar working as Asst Professor in YCOE, Regional campus of Punjabi University, Patiala in the department of Electronics & Communication Engineering. Dr. Kumar is a lifetime member of IMS (Indian Microelectronics Society) and actively working in the field of Microelectronics and Antennas. He has done his B.Tech in Electronics & Communication Engineering from G.N.D.E.C, Ludhiana, Punjab in 2001, M.Tech in Microelectronics from UCIM, Panjab University, Chandigarh in 2005 and Ph.D from SLIET Longowal in 2016.

B. K. Madhavi received Ph.D. from JNTU, Hyderabad. She completed ME from BITS-PILANI in the specialization of 'Microelectronics'. She published 26 research papers in various National and International Journals and Conferences. Presently she is guiding 10 Ph.D. Students and guided several BTech and MTech Projects. She has also reviewed research papers for IETE. She participated in several workshops, summer and winter schools, National, International conferences and also organized several National level workshops and seminars etc. Her research interest includes Microelectronics (VLSI Design, Low Power VLSI, Mixed Signal Processing), Wireless communications.

Girish Mekala joined the Electronics and Communication Engineering department in 2017, where he is currently an Associate Professor in Vidya Jyothi Institute of Technology, Hyderabad. His research area in the field of VLSI, Nanotechnology, 3D IC Packaging system, on-chip interconnects, through silicon vias (TSV). He received his Ph.D. and M.Tech. degrees in VLSI Design Automation and Techniques in E&CE Department from National Institute of Technology, Hamirpur, Himachal Pradesh, India. He is reviewer of several reputed Journals.

Deep Kamal Randhawa is working in the research area of Nanoelectronics. The research work is basically focussed on nanodevices, sensors and thermoelectric effects using nanomaterials using quantum computational methods.

Suman Rani is pursuing her Ph.D. from IKG PTU Jalandhar and she has accomplished her Masters of Technology degree from Guru Jambheshwar University of Science & Technology Hisar. in 2007- 2009. She is currently serving as Assistant Professor in Baba Farid College of Engg. & Technology, Bathinda. She has 10+ years of teaching experience to both undergraduate and postgraduate students. Rani has published the number of papers in the International & National Journal and Conferences. Her current interest comprises VLSI Design & Nanotechnology.

Karmjit Sandha received B.E. and M.Tech degree in Electronics and Communication Engineering in 1999 and 2007; respectively. He received Ph.D. degrees in VLSI Interconnects from Thapar University, Patiala, India in 2015. He has been serving as Assistant Professor in the Department of Electronics and Communication Engineering of Thapar University, Patiala, Punjab, India. He has been actively working in the research field of Carbon Nanotubes (SWCNT and MWCNT bundle) and its applications as VLSI interconnects and supervising the PhD and M.Tech students for their research work. He has published 30 research papers in SCI/SCIE/SCOPUS indexed peer reviewed journals and 11 research papers in International/National Conferences held in India and abroad.

Balwinder Singh has completed his PhD in 2014 from GNDU Amritsar & also has obtained his Bachelor of Technology degree from National Institute of Technology, Jalandhar and Master of Technology degree from University Centre for Inst. & Microelectronics (UCIM), Panjab University, Chandigarh in 2002 and 2004 respectively. He is currently serving as Pr. Engineer in Center for Development of Advanced Computing (CDAC), Mohali and is a part of the teaching faculty. He has

10+ years of teaching experience to both undergraduate and postgraduate students. Singh has published two books and many papers in the International & National Journal and Conferences. His current interest includes Low Power Techniques, VLSI Design & Testing, and System on Chip.

Paramjot Singh is a student, Electronics and Communication Department, Guru Nanak Dev University Regional campus Jalandhar Area of interest- Nanoelectronics and Optical communication.

Rajendra Prasad Somineni is a Professor, Department of Electronics and Communications, Vallurupalli Nageswara Rao Vignana Jyothi Institute of Engineering & Technology (VNR VJIET), Hyderabad, INDIA, Hyderabad. He has Published 26 Research Papers in International/National Journals/Conferences. His areas of interest are Low Power VLSI, CNTFET/GNRFET based VLSI circuit Designs, Nano technology, Embedded Systems, and Microcontrollers.

Index

M

N

O

P

S

T

V

www.ingramcontent.com/pod-product-compliance
Lightning Source LLC
LaVergne TN
LVHW081326300926
849161LV00008B/1885
* 9 7 8 1 7 9 9 8 1 3 9 3 4 *